A-Level Maths for OCR C2

Paul Sanders

Text © Nelson Thornes 2005
Original illustrations © Nelson Thornes Ltd 2005

Published in 2005 by:
Nelson Thornes Ltd
Delta Place
27 Bath Road
CHELTENHAM
GL53 7TH
United Kingdom

05 06 07 08 09 / 10 9 8 7 6 5 4 3 2 1

A catalogue record for this book is available from the British Library

ISBN 0 7487 9454 9

Sample paper written by Val Dixon

Page make-up by Mathematical Composition Setters Ltd, Salisbury, United Kingdom

Printed and bound in Spain by Graphycems

Acknowledgements

We are grateful to Oxford Cambridge and RSA Examination Board for permission to reproduce all the questions marked OCR.
All answers provided for examination questions are the sole responsibility of the author.

The publishers have made every effort to contact copyright holders but apologise if any have been overlooked.

CONTENTS

INTRODUCTION

A-Level Maths for OCR is a brand new series from Nelson Thornes designed to give you the best chance of success in Advanced Level Maths. This book fully covers the OCR **C2** module specification.

In each chapter, you will find a number of key features:

- A beginning of chapter **OBJECTIVES** section, so you can see clearly what you should learn from each chapter

- **WORKED EXAMPLES** taking you through common questions, step by step

- Carefully graded **EXERCISES** to give you thorough practice in all concepts and skills

- Highlighted **KEY POINTS** to help you see at a glance what you need to know for the exam

- An **IT ICON** (IT) to highlight areas where IT software such as Excel may be used

- **EXTENSION** boxes with background information and additional theory

- An end-of-chapter **SUMMARY** to help with your revision

- An end-of-chapter **REVISION EXERCISE** so you can test your understanding of the chapter

At the end of the book, you will find a **MODULE REVISION EXERCISE** containing exam-type questions for the entire module. This is divided into four sections, mirroring the structure of the specification. [Each section tells you which chapters you should have done.]

Finally, there is a **SAMPLE EXAM PAPER** written by an OCR examiner which you can do under timed exam conditions to see just how well prepared you are for the real exam.

1 Trigonometric functions

The purpose of this chapter is to enable you to

- extend the definitions of sine, cosine and tangent to include angles of any magnitude

- sketch and use the graphs of the sine, cosine and tangent functions

- calculate exact values for the trigonometric ratios of 30°, 45° and 60°

- use the basic trigonometric identities

$$\cos^2\theta + \sin^2\theta \equiv 1 \quad \text{and} \quad \tan\theta \equiv \frac{\sin\theta}{\cos\theta}$$

  to prove other identities.

- solve simple trigonometric equations

In earlier work you will certainly have made use of the cosine, sine and tangent functions to solve problems involving right-angled triangles. In this chapter, important properties of these three functions are investigated and developed. This chapter begins by giving definitions for each of the trigonometric functions that apply to angles of any magnitude.

Definitions of Cosine and Sine

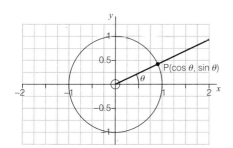

A circle of radius 1 unit and centre the origin is drawn.

A ray, or half line, starting at the origin and making an angle θ with the positive x axis is then drawn. (The normal convention is that moving in an **anticlockwise** direction from the positive x axis gives a **positive angle** and that moving in a **clockwise** direction from the positive x axis gives a **negative angle**.)

The point P is the point of intersection of the circle and the ray.

The cosine of the angle θ, usually written as $\cos\theta$, is defined as the x co-ordinate of the point P.

The sine of the angle θ, usually written as $\sin\theta$, is defined as the y co-ordinate of the point P.

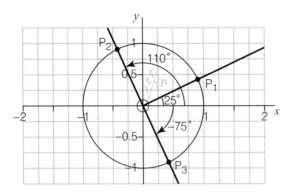

The diagram on the left shows rays coming out from the origin at angles of 25° and 110° and −75°.

The point P_1 is where the ray from the origin that makes an angle of 25° with the x axis meets the circle of centre $(0, 0)$ and radius 1.

The definition tells you that the x co-ordinate of P_1 is the value of cos 25° so, from the graph, you can see that $\cos 25° = x_{P_1} \approx 0.90$.

Similarly, the y co-ordinate of P_1 is the value of sin 25° so, from the graph, you can see that

$$\sin 25° = y_{P_1} \approx 0.40.$$

The point P_2 is where the ray from the origin that makes an angle of 110° with the x axis meets the circle. Using this point it can be said that

$$\cos 110° = x_{P_2} \approx -0.35$$

and

$$\sin 110° = y_{P_2} \approx 0.95.$$

The point P_3 is where the ray from the origin that makes an angle of −75° with the x axis meets the circle. Using this point it can be said that

$$\cos(-75°) = x_{P_3} \approx 0.25$$

and

$$\sin(-75°) = y_{P_3} \approx -0.95.$$

Check that these values for the sine and cosine of 25°, 110° and −75° agree with the values you can obtain from your calculator.

Remember that inaccuracies from reading the graph will mean that only approximate agreement can be expected.

Symmetry Properties of the Cosine and Sine Functions

Knowledge of the values of sine and cosine for a particular angle together with the symmetries of a circle allow the values of the sine and cosine of related angles to be deduced.

EXAMPLE 1

Given that cos 25° = 0.9063, sin 25° = 0.4226 correct to four decimal places, obtain, without using a calculator, the values of

i) cos(−25°)   and   sin(−25°)

ii) cos 155°   and   sin 155°

iii) cos 205°   and   sin 205°

iv) cos 65°   and   sin 65°

EXAMPLE 1 (continued)

S
O
L
U
T
I
O
N

i)

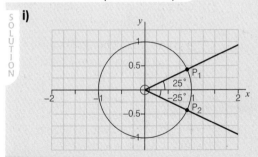

The ray for an angle of $-25°$ is the reflection in the x axis of the ray for $25°$ and, since the circle is certainly symmetrical about the x axis, you can write:

$$\cos(-25°) = x \text{ co-ordinate of } P_2$$
$$= x \text{ co-ordinate of } P_1$$
$$= \cos 25° = 0.9063 \text{ (4 d.p.)}$$

and
$$\sin(-25°) = y \text{ co-ordinate of } P_2$$
$$= -y \text{ co-ordinate of } P_1$$
$$= -\sin 25° = -0.4226 \text{ (4 d.p.)}$$

ii)

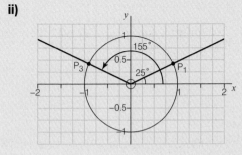

The ray for an angle of $155°$ is the reflection in the y axis of the ray for $25°$ and, since the circle is certainly symmetrical about the y axis, you can write:

$$\cos 155° = x \text{ co-ordinate of } P_3$$
$$= -x \text{ co-ordinate of } P_1$$
$$= -\cos 25° = -0.9063 \text{ (4 d.p.)}$$

and
$$\sin 155° = y \text{ co-ordinate of } P_3$$
$$= y \text{ co-ordinate of } P_1$$
$$= \sin 25° = 0.4226 \text{ (4 d.p.)}$$

iii)

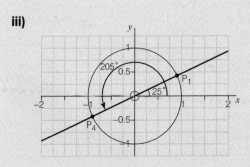

The ray for an angle of $205°$ is the image of the ray for $25°$ after a half turn about the origin so

$$\cos 205° = x \text{ co-ordinate of } P_4$$
$$= -x \text{ co-ordinate of } P_1$$
$$= -\cos 25° = -0.9063 \text{ (4 d.p.)}$$

and
$$\sin 205° = y \text{ co-ordinate of } P_4$$
$$= -y \text{ co-ordinate of } P_1$$
$$= -\sin 25° = -0.4226 \text{ (4 d.p.)}$$

iv)

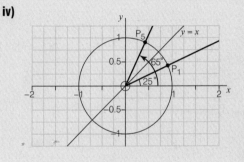

The ray for an angle of $65°$ is the reflection in the line $y = x$ of the ray for $25°$ so

$$\cos 65° = x \text{ co-ordinate of } P_5$$
$$= y \text{ co-ordinate of } P_1$$
$$= \sin 25° = 0.4226 \text{ (4 d.p.)}$$

and
$$\sin 65° = y \text{ co-ordinate of } P_5$$
$$= x \text{ co-ordinate of } P_1$$
$$= \cos 25° = 0.9063 \text{ (4 d.p.)}$$

The symmetry properties used in example 1 can be applied to obtain the general results:

$$\cos(-\theta) = \cos\theta \qquad \cos(180 - \theta) = -\cos\theta \qquad \cos(180 + \theta) = -\cos\theta \qquad \cos(90 - \theta) = \sin\theta$$
$$\sin(-\theta) = -\sin\theta \qquad \sin(180 - \theta) = \sin\theta \qquad \sin(180 + \theta) = -\sin\theta \qquad \sin(90 - \theta) = \cos\theta$$

EXERCISE 1

1 Without using a calculator, write down the values of
a) $\cos 0°$ and $\sin 0°$
b) $\cos 90°$ and $\sin 90°$
c) $\cos 180°$ and $\sin 180°$
d) $\cos 270°$ and $\sin 270°$
e) $\cos 360°$ and $\sin 360°$
f) $\cos(-90°)$ and $\sin(-90°)$
g) $\cos(-180°)$ and $\sin(-180°)$
h) $\cos(-270°)$ and $\sin(-270°)$

2 Without using a calculator, use the facts that
$$\cos 30° = 0.87 \qquad \sin 30° = 0.50 \quad \text{(to 2 d.p.)}$$
and the basic definitions of sin and cos to obtain the values of
a) $\cos(-30°)$ and $\sin(-30°)$
b) $\cos 150°$ and $\sin 150°$
c) $\cos 210°$ and $\sin 210°$
d) $\cos 330°$ and $\sin 330°$
e) $\cos 390°$ and $\sin 390°$
f) $\cos 510°$ and $\sin 510°$
g) $\cos 570°$ and $\sin 570°$
h) $\cos 690°$ and $\sin 690°$
i) $\cos(-150°)$ and $\sin(-150°)$
j) $\cos(-210°)$ and $\sin(-210°)$
k) $\cos(-330°)$ and $\sin(-330°)$

3 Without using a calculator, use the facts that
$$\cos 60° = 0.50 \qquad \sin 60° = 0.87 \quad \text{(to 2 d.p.)}$$
and the basic definitions of sin and cos to obtain the values of
a) $\cos(-60°)$ and $\sin(-60°)$
b) $\cos 120°$ and $\sin 120°$
c) $\cos 240°$ and $\sin 240°$
d) $\cos 300°$ and $\sin 300°$
e) $\cos 420°$ and $\sin 420°$
f) $\cos 480°$ and $\sin 480°$
g) $\cos 600°$ and $\sin 600°$
h) $\cos 660°$ and $\sin 660°$
i) $\cos(-120°)$ and $\sin(-120°)$
j) $\cos(-240°)$ and $\sin(-240°)$
k) $\cos(-300°)$ and $\sin(-300°)$

The Graphs of the Cosine and Sine Functions

Using your calculator, or the definitions and symmetry properties of the trigonometric functions discussed in the last section, the table of results shown below can be obtained:

θ	$\cos\theta$	$\sin\theta$		θ	$\cos\theta$	$\sin\theta$		θ	$\cos\theta$	$\sin\theta$
−360	1	0		0	1	0		360	1	0
−330	0.87	0.5		30	0.87	0.5		390	0.87	0.5
−300	0.5	0.87		60	0.5	0.87		420	0.5	0.87
−270	0	1		90	0	1		450	0	1
−240	−0.5	0.87		120	−0.5	0.87		480	−0.5	0.87
−210	−0.87	0.5		150	−0.87	0.5		510	−0.87	0.5
−180	−1	0		180	−1	0		540	−1	0
−150	−0.87	−0.5		210	−0.87	−0.5		570	−0.87	−0.5
−120	−0.5	−0.87		240	−0.5	−0.87		600	−0.5	−0.87
−90	0	−1		270	0	−1		630	0	−1
−60	0.5	−0.87		300	0.5	−0.87		660	0.5	−0.87
−30	0.87	−0.5		330	0.87	−0.5		690	0.87	−0.5

This table can be used to plot the graphs of the two functions:

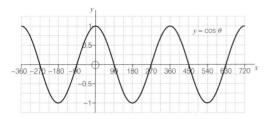

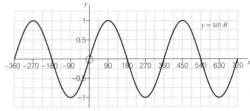

Notice that

- each graph repeats after 360°: it can be said that the cosine function and the sine function are periodic and each have **period 360°**,
- the cosine graph has lines of symmetry at $\theta = 0°$, $\theta = 180°$, $\theta = 360°$, etc.: this is an immediate consequence of the symmetry properties discussed in the last section,
- the sine graph has lines of symmetry at $\theta = 90°$, $\theta = 270°$; $\theta = 450°$, etc.: this is also an immediate consequence of the symmetry properties discussed in the last section.

Definition of Tangent

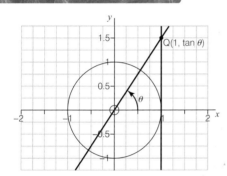

Like the sine and cosine functions, the definition of the tangent function starts with a circle of radius 1 and centre the origin.

The line $x = 1$ is drawn: this line is a tangent to the circle.

A line which passes through the origin and makes an angle θ with the positive x axis is drawn.

The point Q is the point of intersection of this line with the line $x = 1$.

The tangent of the angle θ, usually written as tan θ, is defined as the y co-ordinate of the point Q.

The diagram below shows the line passing through the origin which makes an angle of 30° with the positive x axis. The point Q_1 is where this line intersects the line $x = 1$.

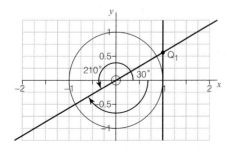

The definition for tangent means that the y co-ordinate of Q_1 is the value of tan 30° and this can be estimated from the graph:

$$\tan 30° = y_{Q_1} \approx 0.6$$

Compare these values with those given by a calculator.

Notice that this diagram also gives the results

$$\tan 210° \approx 0.6$$

and $\tan(-150°) \approx 0.6$.

The next diagram shows the line passing through the origin which makes an angle of $120°$ with the positive x axis. The point Q_2 is where this line intersects the line $x = 1$.

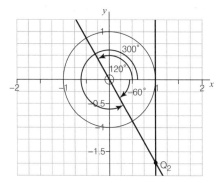

Looking at the y co-ordinate of Q_2 gives the results

$$\tan 120° = \tan 300° = \tan(-60°) \approx -1.7$$

Using similar methods, a table of approximate values for $\tan \theta$ may be produced:

θ	$\tan \theta$		θ	$\tan \theta$		θ	$\tan \theta$
−360	0		0	0		360	0
−330	0.6		30	0.6		390	0.6
−300	1.7		60	1.7		420	1.7
−270	∞		90	∞		450	∞
−240	−1.7		120	−1.7		480	−1.7
−210	−0.6		150	−0.6		510	−0.6
−180	0		180	0		540	0
−150	0.6		210	0.6		570	0.6
−120	1.7		240	1.7		600	1.7
−90	∞		270	∞		630	∞
−60	−1.7		300	−1.7		660	−1.7
−30	−0.6		330	−0.6		690	−0.6

The line that makes an angle of $90°$ with the x axis is parallel to the line $x = 1$ so the lines only meet at infinity.

and this leads to the graph of the tangent function shown below:

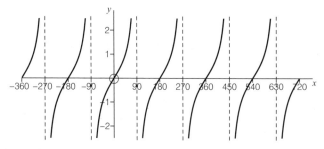

Notice that

- the tan function is periodic and has period $180°$
- the graph has vertical asymptotes at $-90°$, $90°$, $270°$, etc.

An asymptote is a line that the curve gets closer and closer to without touching.

An asymptote is usually shown as a **dotted** line on sketch graphs.

Sketching Related Graphs

The easiest way of sketching graphs such as $y = 3 + \cos x$ and $y = \sin(2x)$ is to recall the work on graphs and transformations from module C1 which is summarised below.

$$y = f(x) \xrightarrow{\text{reflection in } x \text{ axis}} y = -f(x)$$

$$y = f(x) \xrightarrow{\text{translation } \binom{0}{b}} y = f(x) + b$$

$$y = f(x) \xrightarrow{\text{translation } \binom{a}{0}} y = f(x - a)$$

$$y = f(x) \xrightarrow{\text{stretch of scale factor } k \text{ in } y \text{ direction}} y = kf(x)$$

$$y = f(x) \xrightarrow{\text{stretch of scale factor } k \text{ in } x \text{ direction}} y = f\left(\frac{x}{k}\right)$$

$$y = f(x) \xrightarrow{\text{stretch of scale factor } \frac{1}{k} \text{ in } x \text{ direction}} y = f(kx)$$

EXAMPLE 2

Sketch the graph of $y = -\cos x$ for values of x in the interval $-360° \leqslant x \leqslant 360°$.

SOLUTION

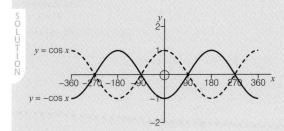

You know that

$$y = \cos x \xrightarrow{\text{reflection in } x \text{ axis}} y = -\cos x$$

so the required graph is the image of $y = \cos x$ after a reflection in the x axis.

EXAMPLE 3

Sketch the graph of $y = 3 + \cos x$ for values of x in the interval $-360° \leqslant x \leqslant 360°$.

SOLUTION

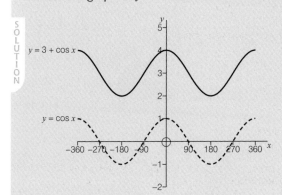

You know that

$$y = \cos x \xrightarrow{\text{translation } \binom{0}{3}} y = \cos x + 3$$

so the required graph is the image of $y = \cos x$ after a translation of $\binom{0}{3}$.

EXAMPLE 4

Sketch the graph of $y = \sin(x - 45)$ for values of x in the interval $-360° \leqslant x \leqslant 360°$.

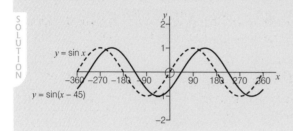

You know that

$$y = \sin x \xrightarrow{\text{translation} \binom{45}{0}} y = \sin(x - 45)$$

so the required graph is the image of $y = \sin x$ after a translation of $\binom{45}{0}$.

EXAMPLE 5

Sketch the graph of $y = \sin 3x$ for values of x in the interval $-360° \leqslant x \leqslant 360°$.

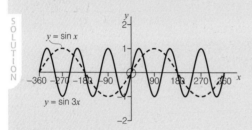

You know that

$$y = \sin x \xrightarrow{\text{stretch of scale factor } \frac{1}{3} \text{ in } x \text{ direction}} y = \sin 3x$$

so the required graph is the image of $y = \sin x$ after a stretch of factor $\dfrac{1}{3}$ in the x direction.

EXERCISE 2

1 On separate diagrams, sketch the graphs of the following functions for values of θ between $-180°$ and $540°$:
 a) $y = \cos \theta$ **b)** $y = 2 + \cos \theta$ **c)** $y = 2 \cos \theta$ **d)** $y = \cos 2\theta$

2 Sketch the graphs of the following functions for values of θ between $-180°$ and $540°$:
 a) $y = \sin \theta$ **b)** $y = 2 + \sin \theta$ **c)** $y = 2 \sin \theta$ **d)** $y = \sin 2\theta$

3 Sketch the graphs of the following functions for values of θ between $-180°$ and $540°$:
 a) $y = \tan \theta$ **b)** $y = 2 + \tan \theta$ **c)** $y = 2 \tan \theta$ **d)** $y = \tan 2\theta$

4 Suggest equations for graphs shown in the diagrams:
 a)

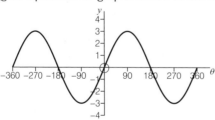

 b)

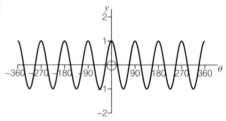

 c)

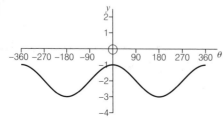

Using Trigonometric Ratios to Solve Right-Angled Triangles

This section will ensure that the new definitions of the trigonometric functions still allow previous methods of solution of right-angled triangle problems to be used.

Consider a right-angled triangle with vertices O, A and B with OB = z, angle OAB = 90° and angle AOB = $\theta°$.

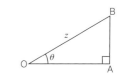

Superimpose the triangle onto the co-ordinate grid with O at the origin and OA lying on the positive x axis.

The ray OB makes an angle θ with the positive x axis.

Using the definitions of cosine and sine, you know that the ray OB meets the circle of radius 1 and centre the origin at the point P(cos θ, sin θ).

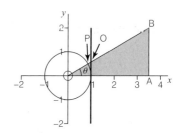

The line segment OB is the image of the line segment OP after an enlargement of scale factor z and centre the origin so the co-ordinates of B are (z cos θ, z sin θ).

The x co-ordinate of B gives the length of OA, or the adjacent side, of the triangle, so you have

$$OA = z \cos \theta$$
$$\Rightarrow \quad \text{adjacent} = \text{hypotenuse} \times \cos \theta$$
$$\Rightarrow \quad \cos \theta = \frac{\text{adjacent}}{\text{hypotenuse}}.$$

Similarly, the y co-ordinate of B gives the length of AB, or the opposite side, of the triangle, so you have

$$AB = z \sin \theta$$
$$\Rightarrow \quad \text{opposite} = \text{hypotenuse} \times \sin \theta$$
$$\Rightarrow \quad \sin \theta = \frac{\text{opposite}}{\text{hypotenuse}}.$$

Using the definition of tangent, you know that the ray OB meets the line x = 1 at the point Q(1, tan θ).

The line segment which joins (1, 0) to Q has length tan θ. The line segment AB is the image of this segment after an enlargement of scale factor OA and centre the origin so the length of AB is given by

$$AB = OA \tan \theta$$
$$\Rightarrow \quad \text{opposite} = \text{adjacent} \times \tan \theta$$
$$\Rightarrow \quad \tan \theta = \frac{\text{opposite}}{\text{adjacent}}.$$

It has been shown that the definitions of the trigonometric functions lead directly onto the three rules used for solving right-angled triangles.

Exact Values of Trigonometric Ratios

Some angles have trigonometric ratios that can be given exactly rather than just approximately. This section will use right-angled triangle methods to calculate the exact values of the trigonometric ratios for 30°, 45° and 60°.

First, consider an equilateral triangle, ABC, whose sides are 2 units long. Let M be the foot of the perpendicular from C to the side AB.

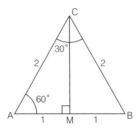

Using Pythagoras's theorem

$$CM^2 = 2^2 - 1^2 = 3$$
$$\implies CM = \sqrt{3}.$$

Consideration of triangle AMC now gives

$$\sin 30° = \frac{\text{opposite}}{\text{hypotenuse}} = \frac{AM}{AC} = \frac{1}{2}$$

$$\cos 30° = \frac{\text{adjacent}}{\text{hypotenuse}} = \frac{CM}{AC} = \frac{\sqrt{3}}{2}$$

$$\tan 30° = \frac{\text{opposite}}{\text{adjacent}} = \frac{AM}{MC} = \frac{1}{\sqrt{3}} = \frac{\sqrt{3}}{3}.$$

> Remember from C1 work on surds:
> $$\frac{1}{\sqrt{3}} = \frac{\sqrt{3}}{\sqrt{3}\sqrt{3}} = \frac{\sqrt{3}}{3}.$$

Consideration of triangle AMC also gives

$$\sin 60° = \frac{CM}{AC} = \frac{\sqrt{3}}{2}, \qquad \cos 60° = \frac{AM}{AC} = \frac{1}{2}, \qquad \tan 60° = \frac{\text{opposite}}{\text{adjacent}} = \frac{MC}{AM} = \frac{\sqrt{3}}{1} = \sqrt{3}.$$

Now consider a right-angled isosceles triangle ABC with AB = BC = 1 unit.

Pythagoras's theorem gives

$$AC^2 = 1^2 + 1^2 = 2$$
$$\implies AC = \sqrt{2}$$

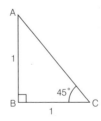

so

$$\sin 45° = \frac{AB}{AC} = \frac{1}{\sqrt{2}} = \frac{\sqrt{2}}{2}$$

$$\cos 45° = \frac{BC}{AC} = \frac{1}{\sqrt{2}} = \frac{\sqrt{2}}{2}$$

and

$$\tan 45° = \frac{BC}{AB} = \frac{1}{1} = 1.$$

Exact values of the trigonometric ratios of other angles can be deduced from these results and the definitions of the ratios.

EXAMPLE 6

Determine the values of sin 120°, cos 120° and tan 120°.

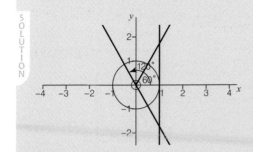

The ray for 120° is the image of the ray for 60° after a reflection in the y axis.

$$\cos 120° = -\cos 60° = -\frac{1}{2}.$$

$$\sin 120° = \cos 60° = \frac{\sqrt{3}}{2}.$$

The line for 120° meets the line $x = 1$ in the same place as the line for −60° meets the line $x = 1$
so tan 120° = tan(−60°) = −tan 60° = −$\sqrt{3}$.

EXERCISE 3

1 Use the results of the previous section together with the basic definitions of the trigonometric functions to write down the **exact** values of

a) sin 150°   b) cos 150°   c) tan 150°   d) sin 210°   e) cos 210°
f) tan 210°   g) sin 240°   h) cos 240°   i) tan 240°   j) sin 300°
k) cos 300°   l) tan 300°   m) sin 330°   n) cos 330°   o) tan 330°

2 Use the results of the previous section together with the basic definitions of the trigonometric functions to write down the **exact** values of

a) sin 135°   b) cos 135°   c) tan 135°   d) sin 225°   e) cos 225°
f) tan 225°   g) sin 315°   h) cos 315°   i) tan 315°

3 Calculate the exact values of
a) $\sin^2 30° + \cos^2 30°$ b) $\sin^2 45° + \cos^2 45°$ c) $\sin^2 180° + \cos^2 180°$

4 Calculate the exact values of

a) $\dfrac{\sin 60°}{\cos 60°}$ b) $\dfrac{\sin 210°}{\cos 210°}$ c) $\dfrac{\sin 180°}{\cos 180°}$

Properties of the Trigonometric Functions

The definitions of cosine, sine and tangent that were met at the beginning of the chapter lead to two very important results that link the values of sin θ, cos θ and tan θ.

Consider again the diagram used to define the trigonometric functions.

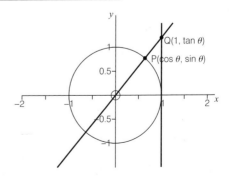

You know P(cos θ, sin θ) is a point on a circle of radius 1 and centre the origin so OP = 1. From C1 work, you also know that the distance between the points (a, b) and (c, d) is $\sqrt{(c-a)^2 + (d-b)^2}$ so you can write the distance between O(0, 0) and P(cos θ, sin θ) as $\sqrt{(\cos\theta)^2 + (\sin\theta)^2}$.

Combining the results gives

$$1 = \sqrt{(\cos\theta)^2 + (\sin\theta)^2}$$
$$\Rightarrow \quad (\cos\theta)^2 + (\sin\theta)^2 = 1.$$

This result is true for **all** angles so you can write

$$(\cos\theta)^2 + (\sin\theta)^2 \equiv 1.$$

> The "≡" sign signifies an **identity** which is a result that is **always** valid. In this case it shows that the result $(\cos\theta)^2 + (\sin\theta)^2 = 1$ is true for all angles.

You would usually write $\cos^2\theta$ as shorthand for $(\cos\theta)^2$ and $\sin^2\theta$ as shorthand for $(\sin\theta)^2$ so the final form for the identity is $\cos^2\theta + \sin^2\theta \equiv 1$.

The points O, P and Q in the diagram lie on a straight line so

$$\text{gradient OP} = \text{gradient OQ}$$
$$\Rightarrow \quad \frac{\sin\theta}{\cos\theta} = \tan\theta.$$

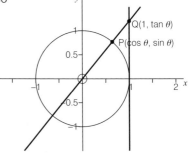

This result is true for any value of θ so you can write

$$\frac{\sin\theta}{\cos\theta} \equiv \tan\theta.$$

Two very **important identities** have been established:

$$\cos^2\theta + \sin^2\theta \equiv 1 \qquad \tan\theta \equiv \frac{\sin\theta}{\cos\theta}$$

> These identities must be learnt.

Calculating Exact Values of Trigonometric Functions

With the aid of these identities, if you know the value of just one of the trigonometric functions then you can calculate the values of the other two functions:

EXAMPLE 7

The angle α is obtuse with $\sin\alpha = \frac{4}{5}$.

Find, without using a calculator, the values of $\cos\alpha$ and $\tan\alpha$.

$$\cos^2\alpha = 1 - \sin^2\alpha = 1 - \frac{16}{25} = \frac{9}{25}$$

$$\Rightarrow \quad \cos\alpha = \pm\frac{3}{5}.$$

EXAMPLE 7 (continued)

Since α is obtuse, cos α is negative

$$\Rightarrow \quad \cos \alpha = -\frac{3}{5}.$$

Moreover

> Referring back to either the definition or the graph of the cosine function you can see that $\cos \theta$ is positive between $0°$ and $90°$, negative from $90°$ to $270°$ and positive again from $270°$ to $360°$.

$$\tan \alpha = \frac{\sin \alpha}{\cos \alpha} = \frac{\frac{4}{5}}{\frac{-3}{5}} = -\frac{4}{3}.$$

EXAMPLE 8

The angle β is acute with $\tan \beta = 3$.

Find the exact values of $\sin \beta$ and $\cos \beta$, giving your answers in simplified surd form.

$$\tan \beta = 3$$

$$\Rightarrow \quad \frac{\sin \beta}{\cos \beta} = 3$$

$$\Rightarrow \quad \sin \beta = 3 \cos \beta$$

$$\Rightarrow \quad \cos^2 \beta + (3 \cos \beta)^2 = 1$$

> Substitute $\sin \beta = 3 \cos \beta$ into $\cos^2 \beta + \sin^2 \beta \equiv 1$.

$$\Rightarrow \quad \cos^2 \beta + 9 \cos^2 \beta = 1$$

$$\Rightarrow \quad 10 \cos^2 \beta = 1$$

$$\Rightarrow \quad \cos^2 \beta = \frac{1}{10}$$

$$\Rightarrow \quad \cos \beta = \pm \frac{1}{\sqrt{10}} = \pm \frac{\sqrt{10}}{10}$$

Since β is acute, cos β is positive

$$\Rightarrow \quad \cos \beta = \frac{\sqrt{10}}{10}$$

and

$$\sin \beta = 3 \cos \beta = \frac{3\sqrt{10}}{10}.$$

Proving Trigonometric Identities

The two basic results, $\cos^2 \theta + \sin^2 \theta \equiv 1$ and $\tan \theta \equiv \dfrac{\sin \theta}{\cos \theta}$, can be used to prove a large number of further identities.

EXAMPLE 9

Prove the identity $\tan^2 \theta - \sin^2 \theta \equiv \tan^2 \theta \sin^2 \theta$.

A good strategy for proving identities is to start with one side of the identity and move logically, one step at a time towards the other side.

$$\tan^2 \theta - \sin^2 \theta \equiv \left(\frac{\sin \theta}{\cos \theta}\right)^2 - \sin^2 \theta \qquad \left(\text{using } \tan \theta \equiv \frac{\sin \theta}{\cos \theta}\right)$$

$$\equiv \frac{\sin^2 \theta}{\cos^2 \theta} - \sin^2 \theta$$

$$\equiv \frac{\sin^2 \theta - \sin^2 \theta \cos^2 \theta}{\cos^2 \theta} \qquad \text{(writing right hand side as a single fraction)}$$

$$\equiv \frac{\sin^2 \theta(1 - \cos^2 \theta)}{\cos^2 \theta} \qquad \text{factorising the numerator}$$

$$\equiv \frac{\sin^2 \theta \sin^2 \theta}{\cos^2 \theta} \qquad \text{(using } \cos^2\theta + \sin^2\theta \equiv 1\text{)}$$

$$\equiv \frac{\sin^2 \theta}{\cos^2 \theta} \sin^2 \theta$$

$$\equiv \tan^2 \theta \sin^2 \theta \qquad \text{as required.}$$

EXERCISE 4

1 The angle β is acute with $\cos \beta = \dfrac{12}{13}$. Find, without using a calculator, the values of $\sin \beta$ and $\tan \beta$.

2 The angle φ is between $180°$ and $270°$ with $\cos \varphi = -\dfrac{63}{65}$. Find, without using the trigonometric functions on your calculator, the values of $\sin \varphi$ and $\tan \varphi$.

3 The angle λ is obtuse with $\sin \lambda = \dfrac{2}{3}$. Find, without using the trigonometric functions on your calculator, the exact values of $\cos \lambda$ and $\tan \lambda$, giving your answers in simplified surd form.

4 Prove the following identities:

a) $(1 - \cos \theta)(1 + \cos \theta) \equiv \sin^2 \theta$

b) $\sin^2 A - \sin^2 B = \cos^2 B - \cos^2 A$

c) $(1 + \cos \theta)^2 + \sin^2 \theta = 2(1 + \cos \theta)$

d) $(1 + \sin \theta + \cos \theta)^2 = 2(1 + \sin \theta)(1 + \cos \theta)$

e) $\dfrac{1}{\tan \theta} + \tan \theta \equiv \dfrac{1}{\sin \theta \cos \theta}$

f) $\dfrac{\sin A}{1 + \cos A} + \dfrac{1 + \cos A}{\sin A} = \dfrac{2}{\sin A}$

You should use a graphical calculator or computer with questions 5–8.

5 Sketch the graphs of the following functions for values of θ between $-180°$ and $540°$:

a) $y = \sin 2\theta$

b) $y = 2 \sin \theta \cos \theta$.

What identity is suggested by these results?

6 Sketch the graph of $y = \cos^2 \theta - \sin^2 \theta$ for values of θ between $-180°$ and $540°$.
Where have you seen this graph before?
What identity is suggested by these results?

7 Sketch the graph of $y = \dfrac{2 \tan \theta}{1 - \tan^2 \theta}$ for values of θ between $-180°$ and $540°$.

Where have you seen this graph before?
What identity is suggested by these results?

8 There is a trigonometric identity that states that

$$\sin 3\theta = a \sin \theta - 4 \sin^3 \theta$$

where a is a positive integer.
Use your graphical calculator to determine the value of a.

> Proofs of the identities suggested by these questions will be met in later modules.

Elementary Trigonometric Equations

Calculators give a rapid means of obtaining accurate values of trigonometric functions. They can also be used to obtain *some* roots of equations involving trigonometric functions. However a **full** solution usually requires both calculator usage and a knowledge of the basic graphs of the trigonometric functions.

EXAMPLE 10

Find the solution of the equation

$$\cos \theta = 0.5 \qquad 0 \leqslant \theta \leqslant 360°.$$

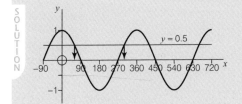

Looking at the cosine graph we can see that there are two roots between 0 and 360°.

Using "INV COS" or "SHIFT COS" on the calculator to obtain the value of $\cos^{-1}(0.5)$ gives one root which is known as the principal value of the equation

$$\cos \theta = 0.5.$$

The calculator gives $\theta = \cos^{-1}(0.5) = 60°$.

Since the graph is symmetrical about the line $x = 180$, the second root can be seen to be 300°. Thus, the full solution of the equation

$$\cos \theta = \frac{1}{2} \qquad 0 \leqslant \theta \leqslant 360°$$

is

$$\theta = 60° \text{ or } 300°.$$

EXAMPLE 11

Find the solution of the equation

$$\sin \theta = -0.5 \qquad -180° < \theta < 540°.$$

EXAMPLE 11 (continued)

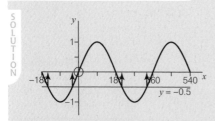

From the graph of $y = \sin x$ you can see that the equation $\sin\theta = -0.5$ has four roots between $-180°$ and $540°$.

Using "INV SIN" or "SHIFT SIN" on the calculator to obtain the value of $\sin^{-1}(-0.5)$ gives a root of the equation

$$\sin\theta = -0.5.$$

The calculator gives $\theta = \sin^{-1}(-0.5) = -30°$.

The sine graph is symmetrical about $x = -90$ so the other negative root is $\theta = -150°$.

The sine graph repeats every $360°$ so there are also roots

at $\qquad \theta = -30° + 360° = 330°$
and at $\qquad \theta = -150° + 360° = 210°$.

Thus the complete solution of the equation

$$\sin\theta = -0.5; \; -180° < \theta < 540°$$

is

$$\theta = -150° \text{ or } -30° \text{ or } 210° \text{ or } 330°.$$

EXAMPLE 12

Find the solution of the equation

$$\tan 2\theta = -1 \qquad 0° < \theta < 180°.$$

Let $z = 2\theta$. Then the equation becomes

$\tan 2\theta = -1 \qquad 0° < \theta < 180°$
$\tan z = -1 \qquad 0° < z < 360°.$

The equation

$$\tan z = -1 \qquad 0° < z < 360°$$

can be solved using the "calculator + graph" approach of the previous two examples.

From the diagram you can see there are two roots in the interval $0° < z < 360°$.

A root of the equation can be found using the $\tan^{-1}$ function on the calculator:

$$z = \tan^{-1}(-1) = -45°.$$

This is **not** one of the required roots since it does not satisfy $0° < z < 360°$.

The two roots can be found from this by using the fact that the tan function repeats every $180°$.

Thus, the roots of $\tan z = -1$ in the given interval will be

$$-45° + 180° = 135°$$

and

$$-45° + 180° + 180° = 315°.$$

EXAMPLE 12 (continued)

The solutions of tan $z = -1$ $0° < z < 360°$ are $z = 135°$ or $315°$.

Finally, you must remember that

$z = 2\theta$

so

$\theta = \dfrac{z}{2} = \dfrac{135}{2}$ or $\dfrac{315}{2} = 67.5°$ or $157.5°$.

The solution of the equation

tan $2\theta = -1$ $0° < \theta < 180°$

is

$\theta = 67.5°$ or $157.5°$.

EXAMPLE 13

Find all the roots of the equation cos $2x = 0.5$ in the interval $-180° < x < 180°$.

cos $2x = 0.5$ $-180° < x < 180°$.

Let $z = 2x$
then the equation   cos $2x = 0.5$ $-180° < x < 180°$
becomes   cos $z = 0.5$ $-360° < z < 360°$.

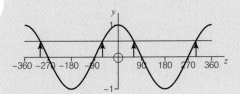

From the graph of $y = \cos z$ it can be seen that there are four solutions to this equation.

Using the calculator gives

$z = \cos^{-1}(0.5)$
$= 60°$.

Using the symmetry of the cosine graph the other solutions are $300°$, $-60°$ and $-300°$.

So cos $z = 0.5$ $-360° < z < 360°$ $\Rightarrow$ $z = -300°, -60°, 60°$ or $300°$.

Finally, you must remember that $z = 2x$ so $x = \dfrac{z}{2} = -150°, -30°, 30°$ or $150°$.

The solution of the equation cos $2x = 0.5$ $-180° < x < 180°$ is

$x = 30°, 150°, -30°,$ or $-150°$.

You could write the solution as
$x = \pm 30°$ or $\pm 150°$.

EXERCISE 5

1 Find all the roots, in the given interval, of the following equations. Give your answers correct to one decimal place.

a) $\cos x = 0.5$ $-360° < x < 720°$

b) $\cos x = -\dfrac{\sqrt{2}}{2}$ $0° < x < 360°$

c) $\cos x = 0.234$ $0° < x < 720°$

d) $\sin x = \dfrac{\sqrt{3}}{2}$ $-360° < x < 360°$

e) $\sin x = -\dfrac{\sqrt{3}}{2}$ $-360° < x < 360°$

f) $\sin x = 0$ $-720° \leqslant x \leqslant 0°$

g) $\tan x = -1.0$ $-360° < x < 360°$

h) $\tan x = 3$ $0° < x < 720°$

i) $\sin x = -0.5$ $0° < x < 180°$

2 Find all the roots, in the given interval, of the following equations:

a) $\cos 3\theta = -1.0$ $-180° \leqslant \theta \leqslant 180°$

b) $\sin 2\theta = 0$ $0° \leqslant \theta \leqslant 360°$

c) $\cos 4\theta = 0$ $0° \leqslant \theta \leqslant 180°$

d) $\sin 3x = 0.5$ $-180° \leqslant x \leqslant 180°$

e) $\tan 3x = 1.0$ $0° \leqslant x \leqslant 180°$

f) $\tan 5x = -\sqrt{3}$ $-90° \leqslant x \leqslant 90°$

g) $\sin 2x = -0.5$ $0° \leqslant x \leqslant 360°$

3 A small particle is attached to a ceiling by means of a spring.

The particle is pulled down and released. t seconds after being released the particle is x cm below the ceiling where $x = 5 + 2\cos(90t)$.

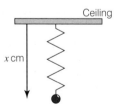

a) Use a graphical calculator to produce a sketch of how x varies with t for $0 \leqslant t \leqslant 8$.

b) Find all the times between 0 and 8 s at which the particle is exactly 6 cm below the ceiling.

Further Trigonometric Equations

You saw earlier that the three trigonometric ratios are linked by the identities

$$\cos^2 \theta + \sin^2 \theta \equiv 1 \qquad \tan \theta \equiv \frac{\sin \theta}{\cos \theta}.$$

These formulae are often used to solve trigonometric equations.

EXAMPLE 14

Solve the equation

$$\tan \theta = 2 \sin \theta \qquad 0° \leqslant \theta \leqslant 360°.$$

SOLUTION

$$\tan \theta = 2 \sin \theta$$

$$\Rightarrow \quad \frac{\sin \theta}{\cos \theta} = 2 \sin \theta$$

$$\Rightarrow \quad \sin \theta = 2 \sin \theta \cos \theta$$

> At this stage it is very important to collect together the terms and **factorise** rather than dividing through by $\sin \theta$.

$$\Rightarrow \quad \sin \theta - 2 \sin \theta \cos \theta = 0$$

$$\Rightarrow \quad \sin \theta (1 - 2 \cos \theta) = 0$$

$$\Rightarrow \quad \sin \theta = 0 \text{ or } \cos \theta = \frac{1}{2}$$

$$\Rightarrow \quad \theta = 0°, 180° \text{ or } 360° \text{ or } \theta = 60° \text{ or } 300°.$$

You can use your graphical calculator to check your solutions to trigonometric equations are correct and to check that you have found all the solutions.

In this case:

- using an x range of 0 to 360 and a y range of −3 to 3, draw the graphs of $y = \tan x$ and $y = 2 \sin x$
- then find the points of intersection of the two graphs.

Remember that solutions found by graphical calculator not supported by the proper working are not acceptable in written assignments.

If you had divided each side of the equation

$$\sin \theta = 2 \sin \theta \cos \theta$$

by $\sin \theta$ you would have obtained

$$1 = 2 \cos \theta$$

$$\Rightarrow \quad \cos \theta = \frac{1}{2}$$

$$\Rightarrow \quad \theta = 60° \text{ or } 300°$$

and you would have lost all the solutions coming from $\sin \theta = 0$.

As a general principle, avoid dividing an equation by something that could be zero; use the alternative of collecting terms together and factorising.

EXAMPLE 15

Find all the roots of the equation $4 \sin^2 \theta + 4 \cos \theta = 5$ in the interval $0 < \theta < 360°$.

$$4 \sin^2 \theta + 4 \cos \theta = 5$$
$$\Rightarrow \quad 4(1 - \cos^2 \theta) + 4 \cos \theta = 5 \qquad \text{since } \sin^2 \theta = 1 - \cos^2 \theta$$
$$\Rightarrow \quad 4 - 4 \cos^2 \theta + 4 \cos \theta = 5$$
$$\Rightarrow \quad 0 = 4 \cos^2 \theta - 4 \cos \theta + 1.$$

This is a quadratic equation in $\cos \theta$. The equation can be solved by factorising:

$$\Rightarrow \quad 0 = (2 \cos \theta - 1)^2$$
$$\Rightarrow \quad \cos \theta = \frac{1}{2}$$
$$\Rightarrow \quad \theta = 60° \text{ or } 300°.$$

EXERCISE 6

1 Solve the equation

$$5 \sin^2 \theta + 3 \cos^2 \theta = 4 \qquad 0 \leqslant \theta \leqslant 360°.$$

2 **a)** Show that the equation $\quad \sin \theta + \cos^2 \theta = 1$
may be written as $\quad \sin \theta - \sin^2 \theta = 0.$
b) Hence find all the roots of the equation $\sin \theta + \cos^2 \theta = 1$ in the interval $0° \leqslant \theta \leqslant 360°$.

3 **a)** Show that the equation $\quad \cos \theta + 3 \sin^2 \theta = 1$
may be written as $\quad 3 \cos^2 \theta - \cos \theta - 2 = 0.$
b) Hence solve the equation $\cos \theta + 3 \sin^2 \theta = 1$ giving values of θ such that $0° \leqslant \theta \leqslant 360°$. Give your answers correct to one decimal place.

4 **a)** Show that the equation $\quad \tan \theta = \sqrt{2} \sin \theta$
may be written as $\quad \sin \theta - \sqrt{2} \sin \theta \cos \theta = 0.$
b) Hence find all values θ such that $0° \leqslant \theta \leqslant 360°$ and $\tan \theta = \sqrt{2} \sin \theta$.

5 Solve the equation $2 \sin^2 \theta = 1 + \sin \theta$ in the interval $0° \leqslant \theta \leqslant 360°$.

6 Solve the equation $6 \cos^2 \theta + \sin \theta - 5 = 0$ in the interval $0° \leqslant \theta \leqslant 360°$, giving your answers correct to one decimal place.

7 Find all the roots of $1 + \cos \theta - 2 \sin^2 \theta = 0$ in the interval $-180° \leqslant \theta \leqslant 180°$.

8 Show that the equation $\quad \tan \theta + 2 \cos \theta = 0$
may be written as $\quad 2 \sin^2 \theta - \sin \theta - 2 = 0.$
Hence find all roots of the equation $\tan \theta + 2 \cos \theta = 0$ in the interval $-180° \leqslant \theta \leqslant 180°$, giving your answers correct to one decimal place.

9 Find all the roots of the equation $3 \sin^2 \theta - \cos^2 \theta = 0$ in the interval $-180° \leqslant \theta \leqslant 180°$.

10 Solve the equation $\tan \theta + 2 \sin \theta = 0$ where θ satisfies $-180° \leqslant \theta \leqslant 180°$.

11 In the diagram the triangles ABC, BCD and DEF are all right-angled. The points CDF lie on a straight line. The points EFG also lie on a straight line.

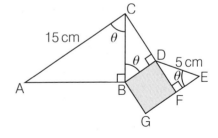

  a) Find expressions involving θ for BC, BD and DF.
  b) If BDFG is a square, prove that θ must satisfy the equation

$$15 \cos^2 \theta = 5 \sin \theta.$$

  c) Solve this equation and hence find the area of the square.

Having studied this chapter you should know how to

● use the circle of radius 1 and centre the origin together with the line $x = 1$ to define $\sin \theta$, $\cos \theta$ and $\tan \theta$
● sketch the graphs of $y = \sin \theta$, $y = \cos \theta$ and $y = \tan \theta$
● obtain and use the exact values

$$\sin 30° = \frac{1}{2}, \cos 30° = \frac{\sqrt{3}}{2}, \tan 30° = \frac{1}{\sqrt{3}} = \frac{\sqrt{3}}{3}$$

$$\sin 45° = \frac{1}{\sqrt{2}} = \frac{\sqrt{2}}{2}, \cos 45° = \frac{1}{\sqrt{2}} = \frac{\sqrt{2}}{2}, \tan 45° = 1$$

$$\sin 60° = \frac{\sqrt{3}}{2}, \cos 60° = \frac{1}{2}, \tan 60° = \sqrt{3}$$

● solve simple equations of the form $\sin k\theta = c$; $\cos k\theta = c$; $\tan k\theta = c$
● use the identities $\cos^2 \theta + \sin^2 \theta \equiv 1$ and $\tan \theta \equiv \dfrac{\sin \theta}{\cos \theta}$ to prove other identities and to solve trigonometric equations

REVISION EXERCISE

1   **a)**   Sketch the graph of $y = \cos x$ for values of x such that $0° \leqslant x \leqslant 720°$.

     b)   On a separate diagram, sketch the graph of $y = \cos(x - 30)$ for values of x such that $0° \leqslant x \leqslant 720°$.

     c)   Solve the equation $\cos(x - 30) = -\dfrac{1}{2}$ for values of x such that $0° \leqslant x \leqslant 720°$.

2   **a)**   Express $2 \sin^2 \theta + 3 \cos \theta$ in terms of $\cos \theta$.

     b)   Hence solve the equation

         $2 \sin^2 \theta + 3 \cos \theta = 3$

     giving all values such that $0° \leqslant \theta \leqslant 360°$.

3   In triangle PQR, shown in the diagram, $\angle Q = 90°$ and $\angle P = \alpha°$. PQ = 7 cm and PR = 8 cm. Find the values of $\sin \alpha°$ and $\tan \alpha°$, giving your answers in surd form.

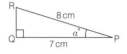

4   **a)**   Prove that $4 \sin^2 \theta - 2 \cos^2 \theta + \tan \theta \cos \theta \equiv 6 \sin^2 \theta + \sin \theta - 2$.

     b)   Hence solve the equation

         $4 \sin^2 \theta - 2 \cos^2 \theta + \tan \theta \cos \theta = 0$

     giving all values such that $0° \leqslant \theta \leqslant 360°$. Where appropriate give answers correct to one decimal place.

5   **a)**   State how the graph of $y = \sin x$ can be transformed onto the graph of $y = \sin 5x$. Hence write down the period of $y = \sin 5x$.

     b)   Solve the equation

         $\sin 5x = \dfrac{\sqrt{3}}{2}$

     giving all values such that $0° \leqslant x \leqslant 90°$.

6   **a)**   Sketch on the same diagram the graphs of $y = \tan x$ and $y = 3 \cos x$ for values of x such that $0° \leqslant x \leqslant 360°$.

     b)   State the number of solutions of the equation $3 \cos x = \tan x$ for values of x such that $0° \leqslant x \leqslant 360°$.

     c)   Show that the equation $3 \cos x = \tan x$ can be rewritten in the form $a \sin^2 x + b \sin x + c = 0$, stating the values of the constants a, b and c.

     d)   Hence solve the equation $3 \cos x = \tan x$ for values of x such that $0° \leqslant x \leqslant 360°$. Where appropriate give answers correct to one decimal place.

7   **a)**   Show that the equation $3 \sin 2\theta - \cos 2\theta = 0$ can be rewritten as $\tan 2\theta = \dfrac{1}{3}$.

     b)   Hence solve the equation $3 \sin 2\theta - \cos 2\theta = 0$, giving all values of θ such that $0° \leqslant \theta \leqslant 360°$. Give your answers correct to the nearest $0.1°$.

                                                 (OCR Jan 2001 P1)

8   **a)**   The angle α is acute and $\tan \alpha = 2$. Find the exact values of $\sin \alpha$ and $\cos \alpha$.

     b)   The angle β is obtuse and $\sin \beta = \dfrac{2}{3}$. Find the exact values of $\cos \beta$ and $\tan \beta$.

2 The Cosine and Sine Rules

The purpose of this chapter is to enable you to

- use the cosine and sine rules in the solution of triangles

In your earlier work you will have certainly used the trigonometric functions to solve problems involving the lengths or angles of right-angled triangles and isosceles triangles. This chapter will extend the techniques used for solving right-angled triangles to techniques that can be used to find the length of unknown sides or size of unknown angles in any triangle.

The Cosine Rule

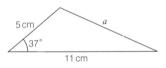

Consider the task of calculating the length a in the triangle shown in the diagram.

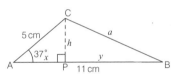

The triangle is not right-angled but it can be split into two right-angled triangles by dropping a perpendicular to the base.

The sides x and h of the triangle ACP can be calculated:

$$\frac{x}{5} = \cos 37° \quad \Rightarrow \quad x = 5 \cos 37° = 3.993 \ldots$$

and

$$\frac{h}{5} = \sin 37° \quad \Rightarrow \quad h = 5 \sin 37° = 3.009 \ldots$$

The length y can be deduced from the calculated value of x since the length AB is known to be 11 cm:

$$y = 11 - x \quad \Rightarrow \quad y = 11 - 3.993 \ldots = 7.006 \ldots$$

The length a can now be found using Pythagoras's theorem:

$$a^2 = 7.006 \ldots^2 + 3.009 \ldots^2 = 58.15 \ldots$$
$$\Rightarrow \quad a = 7.6256 \ldots$$
$$\Rightarrow \quad a = 7.63 \text{ cm} \quad \text{(to two decimal places).}$$

This four-step process could be used to find the length of the third side of a triangle whenever the length of two sides and the angle between these two sides are known. However, the process can be generalised to produce a simple one-step method.

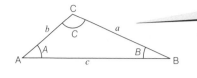

For the triangle with vertices ABC, the usual convention is to use A for the angle at vertex A; B for the angle at vertex B, C for the angle at vertex C; and a for the length BC, b for the length CA and c for the length AB.

Note that length a is **opposite** to angle A, length b is **opposite** to angle B and length c is **opposite** to angle C.

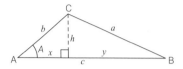

Suppose that the lengths b and c and the angle A of the triangle ABC are already known.

Repeating the argument used in the numerical example, a perpendicular is dropped from C and the lengths x, h and y are determined:

$$\frac{x}{b} = \cos A \quad \Rightarrow \quad x = b \cos A$$

$$\frac{h}{b} = \sin A \quad \Rightarrow \quad h = b \sin A$$

$$y = c - x \quad \Rightarrow \quad y = c - b \cos a$$

a can now be found using Pythagoras' theorem:

$$a^2 = h^2 + y^2$$
$$\Rightarrow \quad a^2 = b^2 \sin^2 A + (c - b \cos A)^2$$
$$\Rightarrow \quad a^2 = b^2 \sin^2 A + c^2 - 2bc \cos A + b^2 \cos^2 A$$
$$\Rightarrow \quad a^2 = b^2 \sin^2 A + b^2 \cos^2 A + c^2 - 2bc \cos A$$
$$\Rightarrow \quad a^2 = b^2 (\sin^2 A + \cos^2 A) + c^2 - 2bc \cos A$$
$$\Rightarrow \quad a^2 = b^2 + c^2 - 2bc \cos A.$$

Remember that $\sin^2 A + \cos^2 A = 1$.

Exactly the same proof will yield $b^2 = c^2 + a^2 - 2ca \cos B$ and $c^2 = a^2 + b^2 - 2ab \cos C$. The three rules are known as the **cosine rule**.

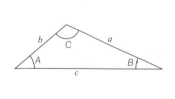

$$a^2 = b^2$$

$$c^2$$

The cosine rule can be used in two circun

- **to work out the third side of a triangle**
  these two sides
- **to work out an angle in a triangle given t**

EXAMPLE 1

A triangle ABC has AB = 10 cm, AC = 8 cm and angle BAC = 37°. Calculate the length of BC.

SOLUTION

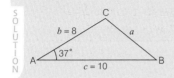

Using the cosine rule gives

$$a^2 = 10^2 + 8^2 - 2 \times 10 \times 8 \times \cos 37° = 36.21 \ldots$$
$$\Rightarrow \quad a = 6.018 \ldots$$
$$\Rightarrow \quad a = 6.02 \text{ cm} \quad \text{(to 3 s.f.)}.$$

EXAMPLE 2

Find the length of the side marked x in the diagram.

> To use the cosine rule to find a side you need to know two sides and the angle between the two known sides.

SOLUTION

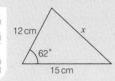

Using the cosine rule gives

$$x^2 = 15^2 + 12^2 - 2 \times 5 \times 12 \times \cos 62° = 199.99 \ldots$$
$$\Rightarrow \quad x = 14.14 \ldots$$
$$\Rightarrow \quad x = 14.1 \text{ cm} \quad \text{(to one decimal place)}.$$

EXAMPLE 3

Find the angle θ in the diagram.

> Take care **not** to interpret this line incorrectly as
> $529 = (256 + 121 - 352)\cos \theta.$

SOLUTION

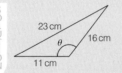

The cosine rule gives

$$23^2 = 16^2 + 11^2 - 2 \times 16 \times 11 \times \cos \theta$$
$$\Rightarrow \quad 529 = 256 + 121 - 352 \cos \theta$$
$$\Rightarrow \quad 352 \cos \theta = 256 + 121 - 529$$
$$\Rightarrow \quad 352 \cos \theta = -152$$
$$\Rightarrow \quad \cos \theta = -\frac{152}{352}$$
$$\Rightarrow \quad \theta = 115.6° \quad \text{(to one decimal place)}.$$

> A useful check at this stage is that the largest angle should be opposite the longest side and the smallest angle should be opposite the shortest side.

EXERCISE 1

Use the cosine rule to find the length of the side marked x or the angle marked θ in the following diagrams, giving your answers correct to four significant figures:

1

2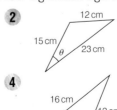

3

4

5 A man walks 5 km on a bearing of 030° to reach a point A. He then walks 8 km due East to reach a point B.
How far is the point B from his starting position?

6 From the top of a 200 m high tower a man can see a cottage, C, on a bearing of 105° and at an angle of depression of 3.2°.
He can also see a hotel, H, on a bearing of 260° and at an angle of depression of 1.3°.
Calculate
a) the distance of the cottage from the foot of the tower;
b) the distance of the hotel from the foot of the tower;
c) the distance of the hotel from the cottage.

7 A triangle has sides of 25 cm, 17 cm and 12 cm. Find the size of the largest angle in the triangle.

8 The distance from a tee to a hole on a golf course is 350 m. If a golfer's drive of 200 m is 180 m short of the hole, find how many degrees his drive was off the direct line.

The Area of a Triangle

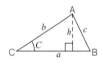

Area of triangle $ABC = \dfrac{1}{2}ah$

but $\dfrac{h}{b} = \sin C$

$\Rightarrow \quad h = b \sin C$

so $\quad \text{area} = \dfrac{1}{2} \times a \times b \sin C = \dfrac{1}{2}ab \sin C.$

This leads to the rule:

$$\text{Area} = \frac{1}{2}ab \sin C$$

Remember:
a and b are two sides of the triangle and C is the angle between those two sides.

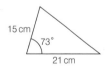

The area of this triangle may be quickly calculated using this formula:

$$\text{area} = \frac{1}{2} \times 21 \times 15 \times \sin 73° = 150.6 \text{ cm}^2 \text{ (to 1 d.p.)}$$

EXERCISE 2

1 Calculate the area of each of the following triangles, giving your answers correct to one decimal place:

a

b

c

2 A school playing field is a quadrilateral FGHI, with a flagpole at the corner F.
G is 127 m from F on a bearing of 35°.
H is 215 m from F on a bearing of 108°.
I is 157m from F on a bearing of 169°.
Calculate the area of the school playing field.

The Sine Rule

You know that:

Area of triangle

$$= \frac{1}{2} ab \sin C = \frac{1}{2} bc \sin A = \frac{1}{2} ca \sin B.$$

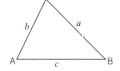

Multiplying through by 2 gives

$$ab \sin C = bc \sin A = ca \sin B.$$

Dividing through by abc gives

$$\frac{ab \sin C}{abc} = \frac{bc \sin A}{abc} = \frac{ac \sin B}{abc}.$$

This simplifies to

$$\frac{\sin C}{c} = \frac{\sin A}{a} = \frac{\sin B}{b}$$

and this equation can be inverted to give

$$\frac{c}{\sin C} = \frac{a}{\sin A} = \frac{b}{\sin B}.$$

These two results are known as the **sine rule**. It can be used whenever the sizes of a side and an opposite angle of the triangle are known.

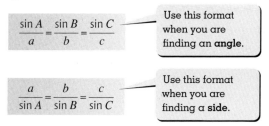

$$\frac{\sin A}{a} = \frac{\sin B}{b} = \frac{\sin C}{c}$$

Use this format when you are finding an **angle**.

$$\frac{a}{\sin A} = \frac{b}{\sin B} = \frac{c}{\sin C}$$

Use this format when you are finding a **side**.

EXAMPLE 4

S
O
L
U
T
I
O
N

Find x and y in the triangle shown.

Using the sine rule gives:

$$\frac{x}{\sin 63°} = \frac{15}{\sin 57°}$$

$$\Rightarrow \quad x = \frac{15}{\sin 57°} \times \sin 63° = 15.936 \ldots$$

$$\Rightarrow \quad x = 15.94 \text{ cm} \quad \text{(to two decimal places).}$$

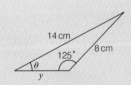

Since the angles of a triangle add to 180°, the third angle of the triangle must be 60° and the sine rule gives

$$\frac{y}{\sin 60°} = \frac{15}{\sin 57°}$$

$$\Rightarrow \quad y = \frac{15}{\sin 57°} \times \sin 60° = 15.489 \ldots$$

$$\Rightarrow \quad y = 15.49 \text{ cm} \quad \text{(to two decimal places).}$$

Care needs to be taken when using the sine rule to find an angle.

It is worth remembering the principles that

- **the largest angle of a triangle is always opposite the longest side**
- **the smallest angle of a triangle is always opposite the shortest side.**

Moreover, it should be remembered that the equation

$$\sin \theta = k \qquad (0 < k < 1)$$

has TWO roots between 0° and 180°.

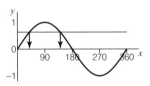

EXAMPLE 5

S
O
L
U
T
I
O
N

Find θ and y in the triangle shown.

$$\frac{\sin \theta}{8} = \frac{\sin 125°}{14}$$

$$\Rightarrow \quad \sin \theta = \frac{\sin 125°}{14} \times 8 = 0.46808 \ldots$$

$$\Rightarrow \quad \theta = 27.910 \ldots$$

$$\Rightarrow \quad \theta = 27.9° \quad \text{(to 1 d.p.).}$$

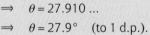

The third angle of the triangle is $180 - (125 + \theta) = 27.0898 \ldots$

The sine rule can now be used to calculate y.

The equation

$$\sin \theta = 0.46808 \ldots \qquad 0 < \theta < 180°$$

has roots

$$\theta = 27.91 \ldots° \text{ or } 152.08 \ldots°.$$

In this case you know θ must be 27.91 ...° since a triangle cannot have two obtuse angles.

EXAMPLE 5 (continued)

$$\frac{y}{\sin 27.0898\ldots°} = \frac{14}{\sin 125°} \implies y = \frac{14}{\sin 125°} \times \sin 27.0898\ldots° = 7.7829\ldots$$

$$\implies y = 7.78 \text{ cm} \quad \text{(to two decimal places)}.$$

EXAMPLE 6

Find the length y and the angle θ in the triangle shown.

Using the cosine rule gives:

$$y^2 = 4^2 + 9^2 - 2 \times 4 \times 9 \cos 19° = 28.922\ldots$$
$$\implies y = 5.3779\ldots$$
$$\implies y = 5.38 \text{ cm} \quad \text{(to two decimal places)}.$$

The sine rule now gives:

$$\frac{\sin \theta}{9} = \frac{\sin 19°}{5.3779\ldots}$$

$$\implies \sin \theta = \frac{\sin 19°}{5.3779\ldots} \times 9 = 0.5448\ldots$$

$$\implies \theta = 146.98\ldots$$

$$\implies \theta = 147.0° \quad \text{(to one decimal place)}.$$

The equation $\sin \theta = 0.5448\ldots$ $0 < \theta < 180°$ has roots $\theta = 33.01\ldots°$ or $146.98\ldots°$.

θ **cannot** be 33° since the longest side of a triangle must be opposite the largest angle and 33.0° cannot possibly be the largest angle in the triangle.

EXERCISE 3

In the following triangles calculate, correct to three significant figures, the values of x, y and θ:

1

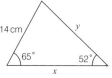

2

3

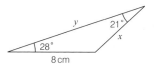

4

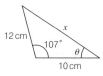

5

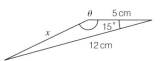

6

7 A triangle has angles in the ratio 3 : 7 : 8 and the longest side of the triangle is 16 cm. John claims that the other two sides must be 6 cm and 14 cm. Do you agree? If not find the lengths of the other two sides.

8 The triangle ABC is such that AB = 8 cm, AC = 4.5 cm and angle ABC is 15°.

> This is an example of the "ambiguous case" of the sine rule. The C2 module test will **not** include examples like this.

Show, by construction, that two different (not congruent) triangles may be drawn to meet these conditions. Measure the length of BC and the two remaining angles in each of these possible triangles.

Use the sine rule to calculate accurately BC and the remaining two angles in each of these possible triangles.

Having studied this chapter you should know how to

- find the area of a triangle using the formula $= \frac{1}{2} ab \sin C$

- solve triangles using the cosine rule $a^2 = b^2 + c^2 - 2bc \cos A$ and the sine rule $\frac{\sin A}{a} = \frac{\sin B}{b} = \frac{\sin C}{c}$ or $\frac{a}{\sin A} = \frac{b}{\sin B} = \frac{c}{\sin C}$

REVISION EXERCISE

Find the values x, y and θ in the following triangles:

1

2

3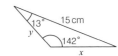

4 Three beacons A, B and C are situated on moor land. A is 1500 m North of B and C is 3200 m from B on a bearing of 320°. What is the distance and bearing of C from A?

5 ABCD is a parallelogram with ∠DAB = 60°, AB = 15 cm, AD = 8 cm. Find the length of the diagonal BD, the size of angle ABD and the area of the parallelogram.

6 A lighthouse, L, is 10 miles due North of the harbour mouth, M. At 1200 hours a ship leaves the harbour and sails on a steady course of 030° with a steady speed. The points P and Q are the two points on the ship's course where the distance of the ship from the lighthouse is 6 miles. The ship will reach the point P before it reaches the point Q.

Draw a sketch diagram to show the points M, L, P and Q.
Calculate the size of angle MQL and the distance MQ.
Calculate the size of angle MPL and the distance MP.

The ship reaches P at 1300 hours. A motor boat is launched from the ship and goes directly to the lighthouse to drop off supplies (the unloading operation takes 10 minutes) and then goes directly to Q where it meets the ship again. Calculate the speed of the motor boat.

7 **a)** Prove that the area of an equilateral triangle whose sides are of length x is

$$\frac{\sqrt{3}}{4}x^2.$$

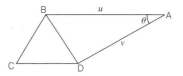

b) The diagram shows a quadrilateral ABCD.
The points BCD form an equilateral triangle.

If $AB = u$; $AD = v$ and $\angle BAD = \theta$, prove that the area of the quadrilateral is

$$\frac{1}{4}(\sqrt{3}u^2 + \sqrt{3}v^2 + 2uv\sin\theta - 2\sqrt{3}uv\cos\theta).$$

8 In the triangle ABC, $AB = 2\sqrt{2}$, $BC = 6\sqrt{2}$ and angle $ABC = 60°$.

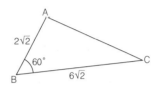

i) Find the length of AC, giving your answer in simplified surd form.

ii) Find the area of triangle ABC, giving your answer in simplified surd form.

(OCR Jan 2002 P1)

9 The diagram shows a quadrilateral ABCD.
$AB = 5$ cm, $BC = 4$ cm, $DA = 3$ cm,
angle DAB = angle $CBD = \theta$ and angle $CDB = 90°$.

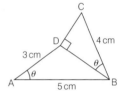

a) By considering the length of BD in two different ways, show that θ must satisfy the equation

$$8\cos^2\theta + 15\cos\theta - 17 = 0.$$

b) Solve this equation and hence determine the area of the quadrilateral.

3 Working with Polynomials

The purpose of this chapter is to enable you to

- divide a polynomial by a linear polynomial
- factorise polynomials
- use the remainder and factor theorems

Polynomial Division

In C1 you saw how to add, subtract or multiply two polynomials. This section considers the division of a polynomial by a polynomial of degree one.

> A polynomial of degree one is a polynomial where the highest power of x appearing is x^1. Examples of polynomials of degree one are $x + 3$ and $5x - 4$.

Consider the problem of dividing $x^2 + 7x - 3$ by $x - 4$, which is the same as simplifying the expression $\dfrac{x^2 + 7x - 3}{x - 4}$.

$$\frac{x^2 + 7x - 3}{x - 4} = \frac{x(x - 4) + 4x + 7x - 3}{x - 4}$$

> Start by trying to write the x^2 term as a multiple of the denominator together with a correcting term.
> You can write $x^2 = x(x - 4) + 4x$.

$$= \frac{x(x - 4) + 11x - 3}{x - 4} = \frac{x(x - 4)}{x - 4} + \frac{11x - 3}{x - 4}$$

$$= x + \frac{11x - 3}{x - 4}$$

> Now write the x term as a multiple of the denominator together with a correcting term:
> You can write $11x = 11(x - 4) + 44$.

$$= x + \frac{11(x - 4) + 44 - 3}{x - 4} = x + \frac{11(x - 4)}{x - 4} + \frac{41}{x - 4}$$

$$= x + 11 + \frac{41}{x - 4}.$$

The result $\dfrac{x^2 + 7x - 3}{x - 4} = x + 11 + \dfrac{41}{x - 4}$

can also be written as $x^2 + 7x - 3 = (x - 4)(x + 11) + 41$.

It is said that

- $(x + 11)$ is the **quotient** when $x^2 + 7x - 3$ is divided by $(x - 4)$
- 41 is the **remainder** when $x^2 + 7x - 3$ is divided by $(x - 4)$.

> Compare these statements with the arithmetic statements
>
> $$\frac{106}{9} = 11\frac{7}{9}, \qquad 106 \div 9 = 11 \text{ remainder } 7,$$
>
> 11 is the quotient when 106 is divided by 9;
> 7 is the remainder when 106 is divided by 9.

EXAMPLE 1

Simplify $\dfrac{x^3 + 4x - 2}{x + 3}$.

> First use the fact that $x^3 = x^2(x+3) - 3x^2$.

$$\frac{x^3 + 4x - 2}{x + 3} = \frac{x^2(x+3) - 3x^2 + 4x - 2}{x + 3}$$

$$= x^2 + \frac{-3x^2 + 4x - 2}{x + 3}$$

> Now use $-3x^2 = -3x(x+3) + 9x$.

$$= x^2 + \frac{-3x(x+3) + 9x + 4x - 2}{x + 3}$$

$$= x^2 - 3x + \frac{13x - 2}{x + 3}$$

> Finally, use $13x = 13(x+3) - 39$

$$= x^2 - 3x + \frac{13(x+3) - 39 - 2}{x + 3}$$

$$= x^2 - 3x + 13 + \frac{-41}{x + 3}.$$

The result $\dfrac{x^3 + 4x - 2}{x + 3} = x^2 - 3x + 13 - \dfrac{41}{x + 3}$ can be rewritten as

$x^3 + 4x - 2 = (x+3)(x^2 - 3x + 13) - 41.$

It can be said that:

- $x^2 - 3x + 13$ is the **quotient** when $x^3 + 4x - 2$ is divided by $(x+3)$
- -41 is the **remainder** when $x^3 + 4x - 2$ is divided by $(x+3)$.

An Alternative Method for Polynomial Division

Notice in the two examples considered in the previous section that **when a polynomial f is divided by a polynomial of degree 1, the quotient is a polynomial of degree one less than the degree of f and the remainder is simply a number.**

This observation provides an alternative method for polynomial divisions.

EXAMPLE 2

Simplify $\dfrac{x^3 + 3x - 5}{x - 2}$.

The observation made above implies that when $x^3 + 3x - 5$ is divided by $x - 2$ the quotient should be a quadratic polynomial, for example $ax^2 + bx + c$ and the remainder should be a number, R:

$$\frac{x^3 + 3x - 5}{x - 2} \equiv ax^2 + bx + c + \frac{R}{x - 2}.$$

EXAMPLE 2 (continued)

Multiplying through by the denominator gives

$$x^3 + 3x - 5 \equiv (ax^2 + bx + c)(x - 2) + R$$
$$\Rightarrow \quad x^3 + 3x - 5 \equiv ax^3 - 2ax^2 + bx^2 - 2bx + cx - 2c + R$$
$$\Rightarrow \quad x^3 + 3x - 5 \equiv ax^3 + (b - 2a)x^2 + (c - 2b)x - 2c + R.$$

The values of a, b, c and R can be found by comparing coefficients:

looking at coefficients of x^3 $\quad \Rightarrow \quad 1 = a$
looking at coefficients of x^2 $\quad \Rightarrow \quad 0 = b - 2a \quad \Rightarrow \quad b = 2$
looking at coefficients of x $\quad \Rightarrow \quad 3 = c - 2b \quad \Rightarrow \quad c = 7$
looking at the constant coefficient $\quad \Rightarrow \quad -5 = -2c + R \quad \Rightarrow \quad R = 9$

so $\dfrac{x^3 + 3x - 5}{x - 2} \equiv x^2 + 2x + 7 + \dfrac{9}{x - 2}.$

EXERCISE 1

> If you have access to a computer algebra package, you could use it to check your answers to this exercise.

In questions 1–12, simplify the polynomial divisions

1 $\dfrac{x^3 + x^2 - 5}{x - 1}$

2 $\dfrac{5x^2 + 8x - 7}{x + 4}$

3 $\dfrac{x^3 + 3x - 4}{x - 2}$

4 $\dfrac{5x + 3}{x + 4}$

5 $\dfrac{2x^2}{x + 3}$

6 $\dfrac{2x - 7}{x + 3}$

7 $\dfrac{10x^2 + 11x + 2}{2x + 3}$

8 $\dfrac{6x^3 + x^2 - 10x + 8}{3x - 1}$

9 $\dfrac{5x^2 + 4}{x - 3}$

10 $\dfrac{2x^3 - 4x - 5}{2x + 1}$

11 $\dfrac{5x^3 - 2x^2 + 5x + 5}{5x - 2}$

12 $\dfrac{6x^3 + 5x^2 + 3x + 1}{2x + 1}$

13 Suppose $p(x) = x^2 + 5x - 2$:
   a) divide $p(x)$ by $x - 2$;
   b) write down the remainder when $p(x)$ is divided by $x - 2$;
   c) find the value of $p(2)$.

14 Suppose $q(x) = 3x^2 - 7x - 26$:
   a) divide $q(x)$ by $x - 3$;
   b) write down the remainder when $q(x)$ is divided by $x - 3$;
   c) find the value of $q(3)$.

15 Suppose $r(x) = 3x^3 + 7$:
   a) divide $r(x)$ by $x + 1$;
   b) write down the remainder when $r(x)$ is divided by $x + 1$;
   c) find the value of $r(-1)$.

> Looking back at questions 13–15 what can you say about the remainder when a polynomial $p(x)$ is divided by $x - a$?

Factorisation of Polynomials

In C1 basic techniques of extracting common factors and factorisation of quadratic expressions were used to produce full factorisations of a variety of polynomials.

EXAMPLE 3

Factorise completely $6x^3 + 15x^2 - 54x$.

> Remember to always first look for a common factor!

The expression has a common factor of $3x$:

$$6x^3 + 15x^2 - 54x \equiv 3x(2x^2 + 5x - 18).$$

The expression $2x^2 + 5x - 18$ is a quadratic expression which you can factorise as $(2x + 9)(x - 2)$. The complete factorisation of the cubic polynomial is therefore

$$6x^3 + 15x^2 - 54x \equiv 3x(2x^2 + 5x - 18) \equiv 3x(2x + 9)(x - 2).$$

EXAMPLE 4

Factorise $5x^6 + 15x^4 - 140x^2$.

The expression has a common factor of $5x^2$:
$$5x^6 + 15x^4 - 140x^2 \equiv 5x^2(x^4 + 3x^2 - 28).$$

Putting $y = x^2$ you can see that the expression $x^4 + 3x^2 - 28$ is basically a quadratic expression: $x^4 + 3x^2 - 28 \equiv y^2 + 3y - 28 = (y + 7)(y - 4) = (x^2 + 7)(x^2 - 4)$ so you now have $5x^6 + 15x^4 - 140x^2 \equiv 5x^2(x^2 + 7)(x^2 - 4)$.

> Remember the **difference of two squares** result $x^2 - a^2 \equiv (x + a)(x - a)$.

Although $x^2 + 7$ will not factorise, $x^2 - 4$ can be factorised as $(x + 2)(x - 2)$. The complete factorisation is therefore
$$5x^6 + 15x^4 - 140x^2 \equiv 5x^2(x^2 + 7)(x + 2)(x - 2).$$

The factorisations of examples 3 and 4 relied on the fact that the expression could be readily reduced to a product of quadratic expressions which could then be factorised using C1 methods. These methods will not work with all polynomials: in most cases if a common factor cannot be found for the polynomial then you should start by searching for a linear factor of the form $(x - a)$.

EXAMPLE 5

Factorise completely $2x^3 - 5x^2 - 19x + 42$.

The expression $2x^3 - 5x^2 - 19x + 42$ has no common factor so, if it does factorise, we might hope for a linear factor of the form $(x - a)$ together with a quadratic factor of the form $(px^2 + qx + r)$, where p, q and r are integers. You would be trying to write

$$2x^3 - 5x^2 - 19x + 42 \equiv (x - a)(px^2 + qx + r).$$

You need to make two observations:

- considering the constant term, you must have $-ac = 42$ and this means that a must be a factor (positive or negative) of 42;
- the equation $2x^3 - 5x^2 - 19x + 42 = 0$ can be rewritten as $(x - a)(px^2 + qx + r) = 0$ so $x = a$ must be a root of the equation $2x^3 - 5x^2 - 19x + 42 = 0$.

EXAMPLE 5 (continued)

You can find a value for a by working systematically through the factors of 42 until you find one that is also a root of $2x^3 - 5x^2 - 19x + 42 = 0$.

> Factors of 42 are $\pm 1, \pm 2, \pm 3, \pm 6, \pm 7, \pm 14, \pm 21, \pm 42$. Start with the numbers closest to 0 since the arithmetic is easier.

Let $f(x) = 2x^3 - 5x^2 - 19x + 42$ then

$$f(1) = 2 \times 1^3 - 5 \times 1^2 - 19 \times 1 + 42 = 2 - 5 - 19 + 42 = 20$$
$$f(-1) = 2 \times (-1)^3 - 5 \times (-1)^2 - 19 \times (-1) + 42 = -2 - 5 + 19 + 42 = 54$$
$$f(2) = 2 \times 2^3 - 5 \times 2^2 - 19 \times 2 + 42 = 16 - 20 - 38 + 42 = 0.$$

You now know $x - 2$ is a factor of $2x^3 - 5x^2 - 19x + 42$.
You can write $2x^3 - 5x^2 - 19x + 42 \equiv (x - 2)(px^2 + qx + r)$ and the values of p, q and r can be found by multiplying out the right-hand side:

$$2x^3 - 5x^2 - 19x + 42 \equiv px^3 + qx^2 + rx - 2px^2 - 2qx - 2r$$
$$\equiv px^3 + (q - 2p)x^2 + (r - 2q)x - 2r.$$

Looking at the constant term gives

$$42 = -2r \implies r = -21.$$

Looking at the x^3 term gives

$$2 = p \implies p = 2.$$

Looking at the x^2 term gives

$$-5 = q - 4 \implies q = -1$$

You now have $2x^3 - 5x^2 - 19x + 42 \equiv (x - 2)(2x^2 - x - 21)$.

The quadratic expression can now be factorised to give the complete factorisation of the cubic polynomial:

$$2x^3 - 5x^2 - 19x + 42 \equiv (x - 2)(2x^2 - x - 21) \equiv (x - 2)(2x - 7)(x + 3).$$

The method of writing $2x^3 - 5x^2 - 19x + 42 \equiv (x - 2)(px^2 + qx + r)$, multiplying out the right hand side and then comparing coefficients to find the values of p, q and r is relatively straightforward but time-consuming. There is a quicker method in which very little needs to be written down:

- Since you know $(x - 2)$ is a factor

$$2x^3 - 5x^2 - 19x + 42 \equiv (x - 2)(\ldots)$$

- The second factor must have $2x^2$ in it to obtain the $2x^3$ term when the two brackets are multiplied out

$$2x^3 - 5x^2 - 19x + 42 \equiv (x - 2)(2x^2 \ldots)$$

- The second factor must have -21 as the constant term to obtain the $+42$ when the two brackets are multiplied out

$$2x^3 - 5x^2 - 19x + 42 \equiv (x - 2)(2x^2 \ldots - 21)$$

- You now have $(x - 2)(2x^2 + qx - 21)$ as the factorisation: when this is multiplied out there will be two x^2 terms, $-4x^2$ and qx^2. These must add together to give $-5x^2$ so the value of q must be -1

$$2x^3 - 5x^2 - 19x + 42 \equiv (x - 2)(2x^2 - x - 21)$$

EXERCISE 2

1 By first extracting a common factor, factorise completely the polynomials
   a) $2x^4 - 10x^3 + 12x^2$ **b)** $5t^5 - 80t^4 + 300t^3$

2 By first putting $x^2 = y$, factorise completely
   a) $x^4 - 21x^2 - 100$ **b)** $4x^4 - 25x^2 + 36$

3 Given that $3x^3 - 2x^2 - 47x + 78 = (x - 3)(px^2 + qx + r)$, determine the values of p, q and r and hence factorise $3x^3 - 2x^2 - 47x + 78$ completely.

4 Given that $(x - 3)$ is a factor of $2x^3 + 13x^2 - 22x - 105$, factorise $2x^3 + 13x^2 - 22x - 105$ completely.

5 Given that $(x + 7)$ is a factor of $4x^3 + 35x^2 + 47x - 14$, factorise $4x^3 + 35x^2 + 47x - 14$ completely.

Factorise fully the following expressions:

6 $4x^3 - 8x^2 - 12x$ **7** $x^4 - 10x^2 + 9$ **8** $y^6 - 5y^4 + 4y^2$

9 $3t^4 - 5t^2 - 28$ **10** $3x^4 - 7x^3 - 10x^2$ **11** $y^4 - 81$

12 $y^4 + y^2 - 20$ **13** $2x^4 - x^3 - 3x^2$ **14** $2x^4 - x^2 - 3$

15 $4y^4 - 13y^2 + 9$ **16** $x^3 + 6x^2 - x - 30$ **17** $x^3 - 13x - 12$

18 $x^3 + 6x^2 - 19x - 84$ **19** $3x^3 - 8x^2 - 31x + 60$ **20** $5x^3 + 6x^2 - 23x + 12$

The Remainder Theorem

If the polynomial $f(x)$ is divided by $(x - a)$ to get quotient $P(x)$ and remainder R, then you can write

$$\frac{f(x)}{x - a} \equiv P(x) + \frac{R}{x - a}$$

or, equivalently, $f(x) \equiv (x - a)P(x) + R$.
Putting $x = a$ gives $f(a) = (a - a)P(a) + R \implies R = f(a)$.

This basic result is known as **the remainder theorem**:

> When the polynomial $f(x)$ is divided by $(x - a)$, the remainder is $f(a)$.

EXAMPLE 6

Find the remainder when $7x^4 + 3x^2 - 25x - 15$ is divided by $x - 2$.

If $f(x) = 7x^4 + 3x^2 - 25x - 15$ then, when $f(x)$ is divided by $x - 2$,

remainder $= f(2) = 7 \times 2^4 + 3 \times 2^2 - 25 \times 2 - 15 = 59$.

EXAMPLE 7

Find the remainder when $7x^4 + 3x^2 - 25x - 15$ is divided by $x + 1$.

If $f(x) = 7x^4 + 3x^2 - 25x - 15$ then when $f(x)$ is divided by $x - (-1)$

remainder $= f(-1) = 7 \times (-1)^4 + 3 \times (-1)^2 - 25 \times (-1) - 15 = 20$.

> Remember $x + 1$ can be written as $x - (-1)$.

You have seen that the remainder theorem gives a quick method of finding the remainder when a polynomial is divided by a polynomial of degree one such as $(x - 2)$ or $(x + 1)$ but it is not yet clear how to quickly find the remainder when a polynomial is divided by a polynomial of degree one such as $(4x - 5)$ or $(2x + 1)$.

The proof of the remainder theorem can be adapted to give the remainder when a polynomial $f(x)$ is divided by $px - q$. If the remainder is R then you can write

$$\frac{f(x)}{px - q} \equiv P(x) + \frac{R}{px - q}$$
$$\Rightarrow \quad f(x) \equiv (px - q)P(x) + R.$$

If you now let $x = \dfrac{q}{p}$ then $px - q = 0$

$$f\left(\frac{q}{p}\right) \equiv 0 \times P\left(\frac{q}{p}\right) + R \quad \Rightarrow \quad R = f\left(\frac{q}{p}\right).$$

So you also have:

> When the polynomial $f(x)$ is divided by $(px - q)$, the remainder is $f\left(\dfrac{q}{p}\right)$.

EXAMPLE 8

Find the remainder when $g(x) = 4x^3 + 7x - 2$ is divided by $2x + 1$.

When $g(x)$ is divided by $2x - (-1)$

remainder $= g\left(-\dfrac{1}{2}\right) = 4 \times \left(-\dfrac{1}{2}\right)^3 + 7 \times \left(-\dfrac{1}{2}\right) - 2 = -6$.

The Factor Theorem

Consider the polynomial $g(x) = (x - 2)(x + 4)(x - 7)$.

It is clear that
$$g(2) = (2 - 2)(2 + 4)(2 - 7) = 0$$
$$g(-4) = (-4 - 2)(-4 + 4)(-4 - 7) = 0$$
$$g(7) = (7 - 2)(7 + 4)(7 - 7) = 0.$$

However, there are no other values of x for which $g(x) = 0$ since

$$g(x) = 0$$
$$\Rightarrow \quad (x - 2)(x + 4)(x - 7) = 0$$
$$\Rightarrow \quad x - 2 = 0 \text{ or } x + 4 = 0 \text{ or } x - 7 = 0$$
$$\Rightarrow \quad x = 2 \text{ or } x = -4 \text{ or } x = 7.$$

> If three quantities multiply together to give 0 then one of the quantities must be 0.

Thus, each factor of $g(x)$ corresponds to a root of the equation $g(x) = 0$.

This observation is now extended to a general polynomial f(x):

If $(x - a)$ is a factor of the polynomial f(x) then

$$f(x) = (x - a)g(x) \qquad \text{for some polynomial g;}$$
$$\Rightarrow \quad f(a) = (a - a)g(a) = 0.$$

It has been shown that **if $(x - a)$ is a factor of the polynomial f(x) then f(a) = 0.**

On the other hand, suppose f(a) = 0.

The remainder theorem tells you that

$$\frac{f(x)}{x - a} = g(x) + \frac{R}{x - a} \qquad \text{where } R = \text{the remainder} = f(a).$$

So, if f(a) = 0 then

$$\frac{f(x)}{x - a} = g(x) + \frac{0}{x - a}$$

$$\frac{f(x)}{x - a} = g(x)$$

$$\Rightarrow \quad f(x) = (x - a)g(x)$$

so $(x - a)$ is a factor of f(x).

It has now been shown that **if f is a polynomial with f(a) = 0 then $(x - a)$ is a factor of f.**

Putting these two results together gives **the factor theorem**:

$(x - a)$ is a factor of the polynomial f(x) if and only if f(a) = 0.

The factor theorem can be written as the statement

"$(x - a)$ is a factor of the polynomial f(x) $\Leftrightarrow$ f(a) = 0"

The double headed arrow $\Leftrightarrow$ provides a shorthand for the **two** statements:

$(x - a)$ is a factor of the polynomial f(x) $\Rightarrow f(a) = 0$

and

$f(a) = 0 \quad \Rightarrow \quad (x - a)$ is a factor of the polynomial f(x).

EXAMPLE 9

Solve the equation $2x^3 - x^2 - 15x + 18 = 0$ given that there is an integer root.

Let $f(x) = 2x^3 - x^2 - 15x + 18$.

If there is an integer solution, a, to the equation $2x^3 - x^2 - 15x + 18 = 0$ then f(a) = 0 and f must have a factor of $(x - a)$. You must be able to write

$$f(x) = 2x^3 - x^2 - 15x + 18 = (x - a)(px^2 + qx + r).$$

Since $-ar = 18$, a must be an integer factor of 18.

The factors of 18 are ±1, ±2, ±3, ±6, ±9 or ±18.

You must search systematically through these values until you find a root:

EXAMPLE 9 (continued)

$$f(1) = 2 \times 1^3 - 1^2 - 15 \times 1 + 18 = 2 - 1 - 15 + 18 = 4$$
$$f(-1) = 2 \times (-1)^3 - (-1)^2 - 15 \times (-1) + 18 = -2 - 1 + 15 + 18 = 30$$
$$f(2) = 2 \times 2^3 - 2^2 - 15 \times 2 + 18 = 16 - 4 - 30 + 18 = 0.$$

$x = 2$ is a root of the equation $f(x) = 0$ so $(x - 2)$ must be a factor of $f(x)$.

You can return to the equation

$$2x^3 - x^2 - 15x + 18 = 0$$
$$\Rightarrow \quad (x - 2)(2x^2 + 3x - 9) = 0$$
$$\Rightarrow \quad (x - 2)(2x - 3)(x + 3) = 0$$
$$\Rightarrow \quad x - 2 = 0 \text{ or } 2x - 3 = 0 \text{ or } x + 3 = 0$$
$$\Rightarrow \quad x = 2 \text{ or } 1.5 \text{ or } -3.$$

> You know that $(x - 2)$ is a factor. The second factor must have $2x^2$ in it to obtain the $2x^3$ term when the two brackets are multiplied out. The second factor must have -9 as the constant term to obtain the $+18$ when the two brackets are multiplied out.
>
> You now have $(x - 2)(2x^2 + bx - 9)$ as the factorisation: when this is multiplied out there will be two x^2 terms, $-4x^2$ and bx^2. These must add together to give $-x^2$ so the value of b must be 3.

EXAMPLE 10

Given that $(x + 1)$ and $(x - 2)$ are factors of $f(x) = x^3 + ax^2 + bx + 10$ find the values of a and b.

Since $x + 1$ is a factor, $x = -1$ is a root of $f(x) = 0$ so $f(-1) = 0$.

$$f(-1) = 0$$
$$\Rightarrow \quad (-1)^3 + a(-1)^2 + b(-1) + 10 = 0$$
$$\Rightarrow \quad -1 + a - b + 10 = 0$$
$$\Rightarrow \quad a - b = -9 \qquad\qquad [1]$$

Since $x - 2$ is a factor, $x = 2$ is a root of $f(x) = 0$ so $f(2) = 0$.

$$f(2) = 0$$
$$\Rightarrow \quad 2^3 + a \times 2^2 + b \times 2 + 10 = 0$$
$$\Rightarrow \quad 8 + 4a + 2b + 10 = 0$$
$$\Rightarrow \quad 4a + 2b = -18$$
$$\Rightarrow \quad 2a + b = -9 \qquad\qquad [2]$$

Equations [1] and [2] give a pair of simultaneous equations which may be solved in the normal way:

> At this stage any method, including use of a graphical calculator, could be used to solve the simultaneous equations.

	a	$-$	b	$=$	-9			
	$2a$	$+$	b	$=$	-9			
[+]	$3a$			$=$	-18	$\Rightarrow$	a	$=$	-6
						$\Rightarrow$	b	$=$	3

EXAMPLE 11

Find a cubic equation with integer coefficients whose solutions are 3, −2 and $\frac{1}{2}$.

The factor theorem implies that if a cubic equation f(x) = 0 has roots 3, −2 and $\frac{1}{2}$ then the cubic must have a factor of $x - 3$, a factor of $x - (-2)$ and a factor of $x - \frac{1}{2}$. Putting these three factors together means that

the equation $(x-3)(x+2)\left(x-\dfrac{1}{2}\right)=0$ has solutions 3, −2, $\frac{1}{2}$

multiplying out $\Rightarrow$ $(x^2-x-6)\left(x-\dfrac{1}{2}\right)=0$ has solutions 3, −2, $\frac{1}{2}$

multiplying out $\Rightarrow$ $x^3-\dfrac{3}{2}x^2-\dfrac{11}{2}x+3=0$ has solutions 3, −2, $\frac{1}{2}$

$\Rightarrow$ $2x^3-3x^2-11x+6=0$ has solutions 3, −2, $\frac{1}{2}$.

EXAMPLE 12

When the polynomial $f(z) = z^3 + az^2 + bz - 10$ is divided by $(z - 1)$ and by $(z + 1)$, the remainders are −10 and −6 respectively.
Find the values of a and b.
Given that the equation f(z) = 0 has one integer root prove that this is the only real root.

Using the remainder theorem:

$$-10 = \text{remainder when f(z) is divided by } (z-1) = f(1)$$
$$\Rightarrow \quad -10 = f(1) = 1^3 + a \times 1^2 + b \times 1 - 10$$
$$\Rightarrow \quad -10 = -9 + a + b$$
$$\Rightarrow \quad a + b = -1 \qquad\qquad [1]$$

and

$$-6 = \text{remainder when f(z) is divided by } (z+1) = f(-1)$$
$$\Rightarrow \quad -6 = f(-1) = (-1)^3 + a \times (-1)^2 + b \times (-1) - 10$$
$$\Rightarrow \quad -6 = -11 + a - b$$
$$\Rightarrow \quad a - b = 5 \qquad\qquad [2]$$

Equations [1] and [2] give a pair of simultaneous equations which may be solved in the normal way:

$$\left.\begin{array}{l} a+b=-1 \\ a-b=5 \end{array}\right\} \Rightarrow \quad a=2, b=-3.$$

You know that the equation $f(z) = z^3 + 2z^2 - 3z - 10 = 0$ has an integer root.
This integer must be a factor of 10 so the possible values of the integer root are ±1, ±2, ±5 and ±10.

Working systematically through these values:
you already know that $\quad f(1) = -10 \quad$ and $\quad f(-1) = -6$;
moving on to 2, we obtain $\quad f(2) = 2^3 + 2 \times 2^2 - 3 \times 2 - 10 = 8 + 8 - 6 - 10 = 0$.

EXAMPLE 12 (continued)

So 2 is a root of the equation $f(z) = z^3 + 2z^2 - 3z - 10 = 0$ which means that $z - 2$ must be a factor of f:

$$z^3 + 2z^2 - 3z - 10 = 0$$
$$\Rightarrow (z - 2)(z^2 + 4z + 5) = 0.$$

You know that $(z - 2)$ is a factor. The second factor must have z^2 in it to obtain the z^3 term when the two brackets are multiplied out. The second factor must have +5 as the constant term to obtain the −10 when the two brackets are multiplied out.

Recall that the discriminant of the quadratic equation $ax^2 + bx + c = 0$ is the number $b^2 - 4ac$ and, if this number is negative then the equation has no real roots.

You now have $(z - 2)(z^2 + bz + 5)$ as the factorisation: when this is multiplied out there will be two z terms, $-2z^2$ and bz^2. These must add together to give $2z^2$ so the value of b must be 4.

The discriminant of the equation $z^2 + 4z + 5 = 0$ is $4^2 - 4 \times 1 \times 5 = -4$ so the equation has no real roots.

$z = 2$ is therefore the only real solution of the equation.

EXERCISE 3

1 Show that 2 is a root of the equation $x^3 - 7x^2 + 4x + 12 = 0$ and hence find all the solutions of this equation.

2 Show that −3 is a root of the equation $y^3 + 3y^2 + 6y + 18 = 0$ and hence factorise the polynomial $f(y) = y^3 + 3y^2 + 6y + 18$.

3 Given that $y + 3$ is a factor of the polynomial $y^3 + y^2 + ay + 36$ find the value of a.

4 Given that $(x - 1)$ and $(x - 4)$ are factors of $f(x) = x^3 + ax^2 + 49x + b$
i) find the values of a and b; **ii)** solve the equation f(x) = 0;
iii) hence, or otherwise, solve the equation $y^6 - 14y^4 + 49y^2 - 36 = 0$.

5 Find the exact solutions of the equations
a) $x^3 + 2x^2 - 17x + 6 = 0$ **b)** $x^3 - 6x^2 + x + 28 = 0$

6 Find a quadratic equation with integer coefficients whose roots are $\frac{1}{2}$ and $-\frac{3}{4}$.

7 Find the remainder when
a) $x^3 - 2x + 4$ is divided by $x - 2$; **b)** $x^5 + 3x^2 + 4$ is divided by $x - 2$;
c) $x^6 + 3x^4 - 2x + 4$ is divided by $x + 3$.

8 The polynomial $p(x) = 2x^3 + ax^2 + bx + 3$ has a remainder of −9 when divided by $x - 2$ and a remainder of −84 when divided by $x + 3$.
i) Find the values of a and b.
ii) Given that the equation p(x) = 0 has a positive integer root, find all solutions of this equation.

9 Find the remainder when
a) $6x^2 + 11x - 2$ is divided by $2x - 1$; **b)** $5x^3 + 13x^2 + 11x - 1$ is divided by $5x + 3$

10 The polynomial $q(x) = x^2(x - 2)^2 + mx - 16$ has a factor of $(x - 4)$.
a) Find the value of m. **b)** Find the remainder when q(x) is divided by $(x + 1)$.

11 The polynomial $r(x) = 2x^3 + ax^2 - 3ax + b$ has a factor of $(x - 3)$ and has a remainder of −60 when it is divided by $(x + 2)$.
Find the values of a and b.

12 Find a cubic equation with integer coefficients whose roots are −3, 2 and $\frac{3}{5}$.

Division of a Polynomial by a Quadratic Polynomial

Similar procedures can be used to divide polynomials by quadratic (or higher degree) polynomials as were used to divide a polynomial by a linear polynomial.

EXAMPLE 13

Simplify $\dfrac{2x^4 - 7x^3 + 5}{x^2 + 2x - 1}$.

First use the fact that
$2x^4 = 2x^2(x^2 + 2x - 1) - 4x^3 + 2x^2$

$$\frac{2x^4 - 7x^3 + 5}{x^2 + 2x - 1} = \frac{2x^2(x^2 + 2x - 1) - 4x^3 + 2x^2 - 7x^3 + 5}{x^2 + 2x - 1}$$

Now use
$-11x^3 = -11x(x^2 + 2x - 1) + 22x^2 - 11x$

$$= 2x^2 + \frac{-11x^3 + 2x^2 + 5}{x^2 + 2x - 1}$$

$$= 2x^2 + \frac{-11x(x^2 + 2x - 1) + 22x^2 - 11x + 2x^2 + 5}{x^2 + 2x - 1}$$

Now use
$24x^2 = 24(x^2 + 2x - 1) - 48x + 24$

$$= 2x^2 - 11x + \frac{24x^2 - 11x + 5}{x^2 + 2x - 1}$$

$$= 2x^2 - 11x + \frac{24(x^2 + 2x - 1) - 48x + 24 - 11x + 5}{x^2 + 2x - 1}$$

$$= 2x^2 - 11x + 24 + \frac{-59x + 29}{x^2 + 2x - 1}.$$

The result $\dfrac{2x^4 - 7x^3 + 5}{x^2 + 2x - 1} = 2x^2 - 11x + 24 + \dfrac{-59x + 29}{x^2 + 2x - 1}$

can be rewritten as $2x^4 - 7x^3 + 5 = (x^2 + 2x - 1)(2x^2 - 11x + 24) + (-59x + 29)$ and it can be said that

- $2x^2 - 11x + 24$ is the **quotient** when $2x^4 - 7x^3 + 5$ is divided by $x^2 + 2x - 1$
- $-59x + 29$ is the **remainder** when $2x^4 - 7x^3 + 5$ is divided by $x^2 + 2x - 1$.

Note that when a polynomial f is divided by a polynomial of degree 2 the quotient is a polynomial of degree two less than the degree of f and the remainder is a polynomial of degree 1.

This observation provides an alternative method for polynomial divisions.

EXAMPLE 14

Simplify $\dfrac{x^3 + 3x - 5}{x^2 - 2x + 3}$.

When $x^3 + 3x - 5$ is divided by $x^2 - 2x + 3$ the quotient should be a polynomial of degree 1, for example $ax + b$ and the remainder should be a polynomial of degree 1, for example $Qx + R$:

$$\frac{x^3 + 3x - 5}{x^2 - 2x + 3} \equiv ax + b + \frac{Qx + R}{x^2 - 2x + 3}.$$

EXAMPLE 14 (continued)

Multiplying through by the denominator gives

$$x^3 + 3x - 5 \equiv (ax + b)(x^2 - 2x + 3) + Qx + R$$
$$\Rightarrow \quad x^3 + 3x - 5 \equiv ax^3 - 2ax^2 + 3ax + bx^2 - 2bx + 3b + Qx + R$$
$$\Rightarrow \quad x^3 + 3x - 5 \equiv ax^3 + (b - 2a)x^2 + (3a - 2b + Q)x + 3b + R.$$

Looking at coefficients of x^3 $\Rightarrow$ $1 = a$

Looking at coefficients of x^2 $\Rightarrow$ $0 = b - 2a$ $\Rightarrow$ $b = 2$

Looking at coefficients of x $\Rightarrow$ $3 = 3a - 2b + Q$ $\Rightarrow$ $Q = 4$

Looking at the constant coefficient $\Rightarrow$ $-5 = 3b + R$ $\Rightarrow$ $R = -11$

So $\dfrac{x^3 + 3x - 5}{x^2 - 2x + 3} \equiv x + 2 + \dfrac{4x - 11}{x^2 - 2x + 3}.$

EXERCISE 4

Simplify the polynomial divisions: **1** $\dfrac{x^3 + 7x}{x^2 + x + 2}$ **2** $\dfrac{3x^4}{x^2 - 1}$

Having studied this chapter you should know

- how to divide a polynomial by a polynomial of degree 1 and be able to identify the resulting quotient and remainder
- that the remainder when a polynomial p(x) is divided by $x - a$ is p(a) (remainder theorem)
- that a polynomial p(x) has a factor $(x - a)$ if and only if p(a) = 0 (factor theorem)

REVISION EXERCISE

1   Given that $(x - 3)$ is a factor of $x^3 + ax^2 + a^2x - 57$ find the possible values of the constant a.

2   **a)**   Given that $(x + 2)$ is a factor of $f(x) = 2x^4 + bx - 20$, find the value of the constant b.
     b)   Find the remainder when $f(x)$ is divided by $(x - 3)$.

3   If $p(x) = 2x^3 + 5x^2 - 2x - 8$
     a)   find the remainder when $p(x)$ is divided by $(x - 2)$.
     b)   find the exact values of the three roots of the equation $p(x) = 0$.

4   Find the quotient and the remainder when
     a) $x^2 + 5x - 2$ is divided by $x + 3$
     b) $2x^3 + 5x - 7$ is divided by $x - 4$
     c) $4x^3 + 9x - 7$ is divided by $2x - 3$.

5 The polynomial $f(x)$ is defined by $f(x) = 2x^3 + bx^2 + 3bx + a$, where a and b are constants.
  i) If $(x + 3)$ is a factor of $f(x)$, find the value of a.
  ii) Find the value of b if the remainder when $f(x)$ is divided by $x - 2$ is -10.

6 Factorise completely
  a) $6x^3 - 13x^2 - 9x + 10$ **b)** $x^4 - 29x^2 + 100$

7 Given that $x - 2$ is a factor of $ax^3 + ax^2 + ax - 42$ find the value of the constant a.

(OCR Nov 2002 P2)

8 The polynomial $p(x)$ defined by $p(x) = x^3 + 5x^2 + ax + 6$ has $x + 3$ as a factor.
  a) Find the value of the constant a.
  b) Find the remainder when $p(x)$ is divided by $x - 4$.
  c) Factorise $p(x)$ completely and hence find the number of real roots of the equation $p(x) = 0$.

9 **i)** Find the remainder when $x^3 - 8x^2 + 11x$ is divided by $(x - 2)$.
  ii) Find the three roots of the equation $x^3 - 8x^2 + 11x + 2 = 0$ giving the two non-integer roots in the exact form $p \pm \sqrt{q}$ where p and q are integers.

(OCR Jun 2001 P2)

10 Solve the equation $2x^3 - 3x^2 - 23x + 12 = 0$.

4 Sequences and Sigma Notation

The purpose of this chapter is to enable you to

- understand and use sequences
- use Σ notation to express sums

Sequences

An Illustration

A building society publishes this table to show the value of a sum of £2000 invested for a period of years in an account that pays 5% per annum compound interest.

Time invested	1 year	2 years	3 years	4 years	5 years
Value of investment	£2100	£2205	£2315.25	£2431.01	£2552.56

The numbers shown in the second row of the table are the first five terms in a **sequence**. If v_n denotes the value of the investment after n years then it is possible to write

$$v_1 = 2100, \qquad v_2 = 2205, \qquad v_3 = 2315.25, \qquad \text{etc.}$$

The terms in the sequence can be generated using the rule $v_n = 2000 \times 1.05^n$ in which the nth term is given as a function of n. Using this rule any term of the sequence can quickly be calculated. For example, $v_{20} = 2000 \times 1.05^{20} = 5306.60$ (2 d.p.) so the value of the investment after 20 years is £5306.60 (to the nearest penny).

Alternatively, the terms in the sequence can be generated from a statement that gives the first term of the sequence together with a method of progressing from one term of the sequence to the next term.

The statements $v_1 = 2100$, $v_{n+1} = 1.05v_n$ express the facts that

- after 1 year the investment is worth £2100;
- the value after $n + 1$ years will be 1.05 times the value after n years, i.e. 105% of the value after n years.

This rule for the sequence can be used to determine the next few terms in the sequence:

$$v_6 = 1.05v_5 = 1.05 \times 2552.56 = 2680.19 \qquad \text{(2 d.p.)}$$
$$v_7 = 1.05v_6 = 1.05 \times 2680.19 = 2814.20 \qquad \text{(2 d.p.)}$$

If v_{20} were to be calculated using this formula, all the intermediate values $v_8, v_9, v_{10}, ..., v_{19}$ would have to be calculated first.

This chapter will continue by first looking at the properties of sequences where the nth term is given as a function of n and then progress to sequences where a starting value and a means of progressing from one term to the next are given.

Sequences of the Form $u_n = f(n)$

A Convergent Sequence

Suppose that the nth term of a sequence $u_1, u_2, u_3, \dots$ is defined by the formula $u_n = \dfrac{2n}{n+1}$.

Then

$$u_1 = \frac{2 \times 1}{1+1} = 1$$

$$u_2 = \frac{2 \times 2}{2+1} = \frac{4}{3}$$

$$u_3 = \frac{2 \times 3}{3+1} = \frac{6}{4}$$

$$u_4 = \frac{2 \times 4}{4+1} = \frac{8}{5}$$

Terms in this sequence may easily be generated on a graphical calculator or on a spreadsheet.

...

$$u_{10} = \frac{20}{11} \approx 1.818 \qquad \text{(3 d.p.)}$$

...

$$u_{100} = \frac{200}{101} \approx 1.980 \qquad \text{(3 d.p.)}$$

...

$$u_{1000} = \frac{2000}{1001} \approx 1.998 \qquad \text{(3 d.p.)}$$

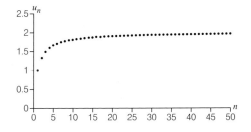

The numbers in the sequence appear to be getting closer and closer to 2. This becomes very apparent if we use a spreadsheet to plot the values of u_n against n.

It can be said that "**the sequence u_n converges to 2 as n tends to infinity**".

We write "**as $n \to \infty$, $u_n \to 2$**" to summarise the fact that as n gets larger and larger the values of u_n get closer and closer to 2.

Polynomial division can be used to give a formal proof of the convergence of this sequence to 2. You start by using polynomial division to obtain an alternative formula for u_n:

$$u_n = \frac{2n}{n+1} = \frac{2(n+1)-2}{n+1} = 2 - \frac{2}{n+1}.$$

Now, as $n \to \infty$, $\dfrac{2}{n+1} \to 0$

so $u_n = 2 - \dfrac{2}{n+1} \to 2 - 0 = 2$ as required.

A Divergent Sequence

Suppose that the nth term of a sequence v_1, v_2, v_3, ... is defined by the formula $v_n = \dfrac{2n^2}{n+1}$.

Then

$$v_1 = \frac{2 \times 1^2}{1+1} = 1$$

$$v_2 = \frac{2 \times 2^2}{2+1} = \frac{8}{3} = 2.6666$$

...

$$v_3 = \frac{2 \times 3^2}{3+1} = \frac{18}{4} = 4.5$$

$$v_4 = \frac{2 \times 4^2}{4+1} = \frac{32}{5} = 6.4$$

...

$$v_{10} = \frac{2 \times 10^2}{10+1} = \frac{200}{11} = 18.18\ldots$$

...

$$v_{100} = \frac{2 \times 100^2}{100+1} = \frac{20\,000}{101} = 198.01\ldots$$

...

$$v_{1000} = \frac{2 \times 1000^2}{1000+1} = \frac{2\,000\,000}{1001} = 1998.00\ldots$$

As n gets larger and larger, it looks as if v_n is also getting larger and larger without reaching a finite limit.

It can be said that "**the sequence v_n is divergent**". We write "**as $n \to \infty$, $v_n \to \infty$**" to summarise the fact that as n gets larger and larger the values of v_n get larger.

> Again polynomial division can be used to formally determine the long time behaviour of the sequence. Using polynomial division it can be shown that
>
> $$v_n = \frac{2n^2}{n+1} = 2n - 2 + \frac{2}{n+1}$$
>
> and from this you can see that as $n \to \infty$, $v_n \to \infty$.

An Oscillating Sequence

Suppose that the nth term of a sequence w_1, w_2, w_3, ... is defined by the formula
$$w_n = \frac{2n}{n+1} \sin(90n^\circ).$$

Then

$$w_1 = \frac{2 \times 1}{1+1} \sin(90 \times 1) = \frac{2}{2} \sin 90^\circ = 1$$

$$w_2 = \frac{2 \times 2}{2+1} \sin(90 \times 2) = \frac{4}{3} \sin 180^\circ = 0$$

$$w_3 = \frac{2 \times 3}{3+1} \sin(90 \times 3) = \frac{6}{4} \sin 270^\circ = -\frac{6}{4} = -1.5$$

$$w_4 = \frac{2 \times 4}{4+1} \sin(90 \times 4) = \frac{8}{5} \sin 360^\circ = 0$$

$$w_5 = \frac{2 \times 5}{5+1} \sin(90 \times 5) = \frac{10}{6} \sin 450^\circ = \frac{10}{6} = 1.66\ldots$$

...

$$w_{100} = \frac{2 \times 100}{100 + 1} \sin(90 \times 100) = \frac{200}{101} \sin 9000° = 0$$

$$w_{101} = \frac{2 \times 101}{101 + 1} \sin(90 \times 101) = \frac{202}{102} \sin 9090° = \frac{202}{102} = 1.9803$$

$$w_{102} = \frac{2 \times 102}{102 + 1} \sin(90 \times 102) = \frac{204}{103} \sin 9180° = 0$$

$$w_{103} = \frac{2 \times 103}{103 + 1} \sin(90 \times 103) = \frac{206}{104} \sin 9270° = -\frac{206}{104} = -1.9807 \ldots$$

...

$$w_{1000} = \frac{2 \times 1000}{1000 + 1} \sin(90 \times 1000) = \frac{2000}{1001} \sin 90\,000° = 0$$

$$w_{1001} = \frac{2 \times 1001}{1001 + 1} \sin(90 \times 1001) = \frac{2002}{1002} \sin 90\,090° = \frac{2002}{1002} = 1.9980 \ldots$$

$$w_{1002} = \frac{2 \times 1002}{1002 + 1} \sin(90 \times 1002) = \frac{2004}{1003} \sin 90\,180° = 0$$

$$w_{1003} = \frac{2 \times 1003}{1003 + 1} \sin(90 \times 1003) = \frac{2006}{1004} \sin 90\,270° = -\frac{2006}{1004} = -1.9980 \ldots$$

...

In this case you can see that the sequence neither converges to a limit nor diverges to $\pm\infty$ but is **oscillating** between values of approximately $+2$, 0 and approximately -2.

The oscillating behaviour of w_n is very apparent if a spreadsheet is used to plot the values of w_n against n.

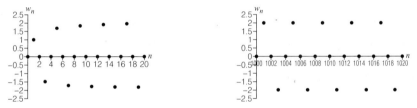

The first graph shows the first 20 numbers in the sequence, whilst the second graph shows the values for w_{1000}, w_{1001}, w_{1002}, ..., w_{1020}.

A Periodic Sequence

Suppose that the nth term of a sequence $x_1, x_2, x_3, \ldots$ is defined by the formula $x_n = \sin(90n°) + 3\cos(180n°)$.

Then

$x_1 = \sin(90°) + 3\cos(180°) = -2$
$x_2 = \sin(180°) + 3\cos(360°) = 3$
$x_3 = \sin(270°) + 3\cos(540°) = -4$
$x_4 = \sin(360°) + 3\cos(720°) = 3$
$x_5 = \sin(450°) + 3\cos(900°) = -2$
$x_6 = \sin(540°) + 3\cos(1080°) = 3$
$x_7 = \sin(630°) + 3\cos(1260°) = -4$
$x_8 = \sin(720°) + 3\cos(1440°) = 3$
$x_9 = \sin(810°) + 3\cos(1620°) = -2$

...

You can see that the sequence repeats through exactly the same four values. It can be said that the sequence has **period** 4.

EXERCISE 1

In each of the sequences in questions 1–12:
a) Find the first five terms of the sequence;
b) Investigate what happens to u_n as $n \to \infty$.

1 $u_n = \dfrac{12}{n}$

2 $u_n = \dfrac{5n - 2}{n + 1}$

3 $u_n = (-1)^n$

4 $u_n = n^2$

5 $u_n = 2^n$

6 $u_n = (1.02)^n$

7 $u_n = (0.9)^n$

8 $u_n = (-0.6)^n$

9 $u_n = (-2.1)^n$

10 $u_n = 3 \cos(45n°)$

11 $u_n = 2 + 5n$

12 $u_n = \dfrac{(-1)^n n}{n + 2}$

13 Which of the following sequences converge to 0 as $n \to \infty$?
a) $u_n = 3^n$ **b)** $u_n = (1.5)^n$ **c)** $u_n = (0.8)^n$
d) $u_n = (0.2)^n$ **e)** $u_n = (-0.6)^n$ **f)** $u_n = (-2.8)^n$

What can be said about the number a if the sequence $u_n = a^n$ converges to 0 as $n \to \infty$?

14 Which of the following sequences converge to 0 as $n \to \infty$?
a) $v_n = n^3$ **b)** $v_n = n^{0.5}$ **c)** $v_n = n^0$
d) $v_n = n^{-0.5}$ **e)** $v_n = n^{-2}$

A sequence is given by the formula $x_n = n^\beta$.
For what values of β will the sequence converge to zero as $n \to \infty$?
For what values of β does the sequence diverge as $n \to \infty$?

15 Determine the limit as $n \to \infty$ of each of the following sequences and use polynomial division to prove your result:
i) $a_n = \dfrac{7n - 5}{n + 4}$ **ii)** $a_n = \dfrac{20 - 2n}{n + 1}$

16 A ball is dropped from a height of 2 metres onto a concrete floor. The maximum height that the ball reaches immediately after its nth impact with the floor is h_n metres where

$$h_n = 2 \times 0.8^n.$$

a) Calculate the values of h_1, h_2, h_3, h_4 and h_5.
b) What happens to h_n as $n \to \infty$?

17 Saffron's grandma put some money into a piggy bank on the day she was born and then each month adds a constant amount of money to the piggy bank.
The amount of money in the piggy bank n months after Saffron was born is £x_n where

$$x_n = 10 + 3n.$$

a) How much money did grandma put in the piggy bank on the day Saffron was born?
b) How much does grandma put into the piggy bank each month?
c) How old will Saffron be when the amount in the piggy bank first exceeds £200?

Sequences where u_{n+1} is Given in Terms of u_n

The sequence $u_1, u_2, u_3, \ldots$ is defined by

$$u_1 = 2, \quad u_{n+1} = 3u_n.$$

The first few terms of the sequence can be easily calculated:

You could interpret the formula $u_{n+1} = 3u_n$ as "new value = 3 × old value"

so

second value = 3 × first value = 3 × 2 = 6
third value = 3 × second value = 3 × 6 = 18.

$u_1 = 2$
$u_2 = 3u_1 = 3 \times 2 = 6$
$u_3 = 3u_2 = 3 \times 6 = 18$
$u_4 = 3u_3 = 3 \times 18 = 54$
$u_5 = 3u_4 = 3 \times 54 = 162$
...

The sequence is **divergent**. As $n \rightarrow \infty$, $u_n \rightarrow \infty$ since each term is three times the previous term.

Now consider the sequence $a_1, a_2, a_3, \ldots$ defined by

$$a_1 = 40, \quad a_{n+1} = 1.2a_n + 15.$$

On a **scientific calculator**, terms of this sequence can be generated by

1) entering the first term (40) and then pressing $\boxed{=}$
This places the first term into the "answer" memory of the calculator.
2) Enter the calculation of the second term as

$$1.2 \; \boxed{\times} \; \boxed{\text{ANS}} \; \boxed{+} \; 15 \; \boxed{=}$$

3) Subsequent terms of the sequence are obtained by just pressing $\boxed{=}$

The terms of the sequence can also be generated on a graphical calculator or on a spreadsheet.

$a_1 = 40$
$a_2 = 1.2a_1 + 15 = 1.2 \times 40 + 15 = 63$
$a_3 = 1.2a_2 + 15 = 1.2 \times 63 + 15 = 90.6$
$a_4 = 123.72$
$a_5 = 163.4 \ldots$
$a_6 = 211.1 \ldots$
$a_7 = 268.3 \ldots$

The terms in this sequence are getting larger and larger and it looks as if the sequence is divergent.
Looking back at the rule we can see that each term is more than 1.2 times the previous term so as $n \rightarrow \infty$, $a_n \rightarrow \infty$.

Consider the sequence given by the rule

$$b_1 = 40, \quad b_{n+1} = 0.9b_n + 15.$$

Again, terms in this sequence can easily be generated on a scientific or graphical calculator or on a spreadsheet.

Then

$$b_1 = 40$$
$$b_2 = 0.9b_1 + 15 = 0.9 \times 40 + 15 = 36 + 15 = 51$$
$$b_3 = 0.9b_2 + 15 = 0.9 \times 51 + 15 = 45.9 + 15 = 60.9$$

...

$$b_{29} = 144.24 \ldots$$
$$b_{30} = 144.81 \ldots$$
$$b_{31} = 145.34 \ldots$$

The values seem to be converging but it is not clear to what value.

...

$$b_{99} = 149.9964 \ldots$$
$$b_{100} = 149.9968 \ldots$$
$$b_{101} = 149.9971 \ldots$$

The sequence appears to converge to 150.

...

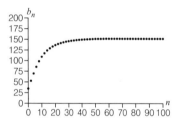

The convergence of the sequence can be illustrated in a graph showing values of b_n against n.

The limiting value could be predicted from the sequence rule by using the following argument.

Suppose

$$a_n \to L \quad \text{as} \quad n \to \infty$$

then you also have

$$a_{n+1} \to L \quad \text{as} \quad n \to \infty.$$

If you allow $n \to \infty$, then the equation

$$a_{n+1} = 0.9a_n + 15$$

becomes

$$L = 0.9L + 15$$
$$\Rightarrow \quad 0.1L = 15$$
$$\Rightarrow \quad L = 150.$$

Care must be taken in using this argument: it does not ensure the sequence does converge, it simply tells us the value the sequence would converge to if it is convergent.

EXAMPLE 1

Find, correct to four decimal places, the first five terms in the sequence given by

$$a_1 = 4, \quad a_{n+1} = \frac{8}{a_n + 2}.$$

Given that the sequence converges to L, find the exact value of L.

SOLUTION

$a_1 = 4$

$a_2 = \dfrac{8}{4 + 2} = 1.33333\ldots = 1.3333$ (to 4 d.p.)

$a_3 = \dfrac{8}{1.33333\ldots + 2} = 2.4$

$a_4 = \dfrac{8}{2.4 + 2} = 1.81818\ldots = 1.8182$ (to 4 d.p.)

$a_5 = \dfrac{8}{1.81818\ldots + 2} = 2.09523\ldots = 2.0952$ (to 4 d.p.)

You are told that

$$a_n \to L \quad \text{as} \quad n \to \infty$$

and you also have

$$a_{n+1} \to L \quad \text{as} \quad n \to \infty.$$

If you allow $n \to \infty$, then the equation $a_{n+1} = \dfrac{8}{a_n + 2}$

becomes

$$L = \frac{8}{L + 2}$$
$$\Rightarrow \quad L^2 + 2L = 8$$
$$\Rightarrow \quad L^2 + 2L - 8 = 0$$
$$\Rightarrow \quad (L + 4)(L - 2) = 0$$
$$\Rightarrow \quad L = -4 \text{ or } 2.$$

The definition

$$a_1 = 4, \quad a_{n+1} = \frac{8}{a_n + 2}$$

implies that all the terms in the sequence must be positive and hence the limit must be positive.

$$\Rightarrow \quad L = 2.$$

EXAMPLE 2

The sequence $u_1, u_2, u_3, ...,$ where u_1 is a given real number, is defined for $n \geqslant 1$ by

$$u_{n+1} = \sqrt{u_n^2} - 4.$$

i) Given that $u_1 = 1$, evaluate u_2, u_3, u_4 and u_5, and describe the behaviour of the sequence.

ii) State the value of u_1 for which all terms in the sequence are equal.

iii) Given that $u_1 = 200$, determine the number of positive terms of the sequence.

SOLUTION

i) $u_1 = 1$

$u_2 = \sqrt{u_1^2} - 4 = \sqrt{1} - 4 = -3$

$u_3 = \sqrt{u_2^2} - 4 = \sqrt{9} - 4 = 5$

$u_4 = \sqrt{u_3^2} - 4 = \sqrt{25} - 4 = 1$

$u_5 = \sqrt{u_4^2} - 4 = \sqrt{1} - 4 = -3$

The sequence will continue by taking the three values 5, 1 and -3 in turn. The sequence has period 3 and can be illustrated by the diagram:

$$1 \longrightarrow -3 \longrightarrow 5$$

ii) If all the values in the sequence are to be equal then $u_2 = u_1$.

$\Rightarrow \quad \sqrt{u_1^2} - 4 = u_1$

$\Rightarrow \quad \sqrt{u_1^2} = u_1 + 4$

$\Rightarrow \quad u_1^2 = (u_1 + 4)^2$

$\Rightarrow \quad u_1^2 = u_1^2 + 8u_1 + 16$

$\Rightarrow \quad 0 = 8u_1 + 16$

$\Rightarrow \quad u_1 = -2$

iii) If $u_1 = 200$ then the first few terms of the sequence are

$u_1 = 200$

$u_2 = \sqrt{200^2} - 4 = 200 - 4 = 196$

$u_3 = \sqrt{196^2} - 4 = 196 - 4 = 192$

$u_4 = \sqrt{192^2} - 4 = 192 - 4 = 188$

You must make the observation that if u_n is non-negative then $\sqrt{u_n^2} = u_n$ so $u_{n+1} = u_n - 4$.

Now

$u_2 = u_1 - 4 = 200 - 4$

$u_3 = u_2 - 4 = 200 - 2 \times 4$

$u_4 = u_3 - 4 = 200 - 3 \times 4$ — Continuing the pattern gives

...

$u_{50} = 200 - 49 \times 4 = 4$

$u_{51} = 200 - 50 \times 4 = 0$

EXAMPLE 2 (continued)

You can see that the first 50 terms of the sequence are positive.

Now $u_{52} = \sqrt{0^2 - 4} = -4$ and $u_{53} = \sqrt{(-4)^2 - 4} = 0$ so all the remaining terms just alternate between 0 and -4.

The sequence $u_1 = 200$, $u_{n+1} = \sqrt{u_n^2 - 4}$ therefore has 50 positive terms.

EXERCISE 2

Each of the sequences in questions 1 to 5 converges. Find the first five terms of each sequence and the exact value of the limit of each sequence.

1 $a_1 = 5$, $a_{n+1} = 0.6a_n + 15$

2 $a_1 = 8$, $a_{n+1} = 30 - 0.2a_n$

3 $a_1 = 2$, $a_{n+1} = 0.95a_n + 10$

4 $a_1 = 2$, $a_{n+1} = \dfrac{12}{a_n + 1}$

5 $a_1 = 4$, $a_{n+1} = \dfrac{6}{a_n + 2}$

Determine whether the sequences in questions 6 to 9 are convergent or divergent. If they are convergent, determine the exact value of the limit.

6 $a_1 = 5$, $a_{n+1} = 0.8a_n + 15$

7 $a_1 = 5$, $a_{n+1} = 1.6a_n + 8$

8 $a_1 = 2$, $a_{n+1} = -5a_n + 15$

9 $a_1 = 5$, $a_{n+1} = 10 - 0.6a_n$

10 The sequence u_1, u_2, u_3, ..., where u_1 is a given real number, is defined for $n \geqslant 1$ by
$$u_{n+1} = \frac{12}{u_n}.$$

  i) Given that $u_1 = 2$, evaluate u_2, u_3, u_4 and u_5, and describe the behaviour of the sequence.

  ii) State the possible values of u_1 for which all terms in the sequence are equal.

11 A model for the number of trees in a forest states that
- the forest currently contains 100 000 trees,
- each year 8% of the trees will be cut down,
- each year 6000 new trees will be planted.

Let t_n denote the number of trees in the forest in n years' time.

a) Explain why t_n satisfies

$$t_0 = 100\,000, \quad t_{n+1} = 0.92t_n + 6000$$

b) Determine the number of trees there will be in the forest in 5 years' time.

c) What does this model predict is going to happen to the number of trees in the forest in the long-term future?

12 A country has a stable population of 60 000 000 people.
The country can be divided into two regions — a prosperous North and a poorer South.

Currently 20 000 000 live in the North whilst 40 000 000 live in the South.

A model for population movement predicts that each year

  8% of the people living in the South move to the North

and

  2% of the people living in the North move to the South.

Let p_n denote the population of the Northern part of the country in n years' time.

a) Prove that p_n satisfies

$$p_0 = 20\,000\,000, \quad p_{n+1} = 4\,800\,000 + 0.9p_n$$

b) Determine predictions for the number of people living in each of the two regions in 3 years' time.

c) What does this model predict is going to happen to the population of the two regions in the long-term future?

Sigma Notation

Many branches of mathematics often make use of the **sum** of a whole series of terms and it is useful to have a shorthand for such expressions so that you don't have to write out every term in the sum.

The shorthand used employs the symbol Σ which is the Greek capital letter S known as sigma. **Write Σ as shorthand for sum**.

The expression $\displaystyle\sum_{r=2}^{7} r^2$

reads as "the sum of r^2 for integer values of r between 2 and 7" so you can write

$$\sum_{r=2}^{7} r^2 = 2^2 + 3^2 + 4^2 + 5^2 + 6^2 + 7^2.$$

> Notice that
>
> $$\sum_{r=1}^{6} (r+1)^2 = (1+1)^2 + (2+1)^2 + \cdots + (6+1)^2$$
>
> $$= 2^2 + 3^2 + 4^2 + 5^2 + 6^2 + 7^2.$$
>
> There are lots of different possible ways of writing the same sum using sigma notation.

Similarly $\displaystyle\sum_{r=1}^{4} 5 \times 1.2^r = 5 \times 1.2^1 + 5 \times 1.2^2 + 5 \times 1.2^3 + 5 \times 1.2^4 = 32.208$

and

$$\sum_{r=1}^{100} \frac{1}{r^2} \text{ is shorthand for } \frac{1}{1^2} + \frac{1}{2^2} + \cdots + \frac{1}{100^2}.$$

> Your graphical calculator will probably allow you to use sigma notation to evaluate sums and obtain
>
> $$\sum_{r=1}^{100} \frac{1}{r^2} = 1.635 \qquad \text{(3 d.p.)}$$

If you have a sequence of terms $a_1, a_2, a_3, \ldots, a_n$ then

$$\sum_{r=1}^{n} a_r \text{ is shorthand for } a_1 + a_2 + a_3 + \cdots + a_n$$

and

$$\sum_{r=1}^{n} r^2 a_r \text{ is shorthand for } 1^2 \times a_1 + 2^2 \times a_2 + 3^2 \times a_3 + \cdots + n^2 \times a_n.$$

Properties of Sigma Notation

There are three simple properties of sigma notation that you will need in the future.

Suppose that

$$a_1 = 3, a_2 = 5 \text{ and } a_3 = 10,$$
$$b_1 = 2, b_2 = 3 \text{ and } b_3 = 6,$$
$$c = 8$$

then

$$\sum_{i=1}^{3} a_i = 3 + 5 + 10 = 18$$

$$\sum_{i=1}^{3} b_i = 2 + 3 + 6 = 11$$

and

$$\sum_{i=1}^{3} (a_i + b_i) = (a_1 + b_1) + (a_2 + b_2) + (a_3 + b_3)$$
$$= (3 + 2) + (5 + 3) + (10 + 6)$$
$$= 5 + 8 + 16$$
$$= 29$$
$$= 18 + 11$$
$$= \sum_{i=1}^{3} a_i + \sum_{i=1}^{3} b_i$$

This result can be generalised to give

Property 1: $\displaystyle\sum_{r=1}^{n} (a_r + b_r) = \sum_{r=1}^{n} a_r + \sum_{r=1}^{n} b_r$

Proof:

$$\sum_{r=1}^{n} (a_r + b_r) = (a_1 + b_1) + (a_2 + b_2) + (a_3 + b_3) + \cdots + (a_n + b_n)$$
$$= (a_1 + a_2 + a_3 + \cdots + a_n) + (b_1 + b_2 + b_3 + \cdots + b_n)$$
$$= \sum_{r=1}^{n} a_r + \sum_{r=1}^{n} b_r.$$

Returning to the example where

$$a_1 = 3, a_2 = 5 \text{ and } a_3 = 10,$$
$$c = 8$$

you have

$$\sum_{i=1}^{3} (ca_i) = ca_1 + ca_2 + ca_3$$

$$= 8 \times 3 + 8 \times 5 + 8 \times 10$$

$$= 24 + 40 + 80$$

$$= 144$$

$$= 8 \times 18$$

$$= c \times \sum_{i=1}^{3} a_i.$$

This result can also be generalised to give

Property 2: If c is a constant then $\displaystyle\sum_{r=1}^{n} (ca_r) = c \sum_{r=1}^{n} a_r$

Proof:

$$\sum_{r=1}^{n} (ca_r) = ca_1 + ca_2 + ca_3 + \cdots + ca_n$$

$$= c(a_1 + a_2 + a_3 + \cdots + a_n)$$

$$= c \sum_{r=1}^{n} a_r.$$

Finally consider $\displaystyle\sum_{i=1}^{3} 8 = 8 + 8 + 8 = 24 = 3 \times 8.$
This can be generalised to give

Property 3: If c is a constant then $\displaystyle\sum_{r=1}^{n} c = nc$

Proof:

$$\sum_{r=1}^{n} c = c + c + c + \cdots + c$$

$$= nc.$$

EXAMPLE 3

Use the information that $\sum_{r=1}^{10} r^2 = 385$ and $\sum_{r=1}^{10} r^3 = 3025$ to deduce the values of

a) $\sum_{r=1}^{10} (r^2 + r^3)$ **b)** $\sum_{r=1}^{10} (3r^2)$ **c)** $\sum_{r=1}^{10} (r^2 + 5)$

a) Using property 1

$$\sum_{r=1}^{10} (r^2 + r^3) = \sum_{r=1}^{10} r^2 + \sum_{r=1}^{10} r^3 = 385 + 3025 = 3410$$

b) Using property 2

$$\sum_{r=1}^{10} (3r^2) = 3 \sum_{r=1}^{10} r^2 = 3 \times 385 = 1155$$

c) Using properties 1 and 3

$$\sum_{r=1}^{10} (r^2 + 5) = \sum_{r=1}^{10} r^2 + \sum_{r=1}^{10} 5 = 385 + 10 \times 5 = 435$$

EXERCISE 3

Use sigma notation to write down shorthand versions of each of the following expressions:

1 $5^3 + 6^3 + 7^3 + \cdots + 20^3$

2 $2^6 + 2^7 + 2^8 + \cdots + 2^{25}$

3 $\dfrac{4}{7} + \dfrac{4}{8} + \dfrac{4}{9} + \dfrac{4}{10} + \cdots + \dfrac{4}{73}$

4 $\dfrac{4}{3^2} + \dfrac{5}{4^2} + \dfrac{6}{5^2} + \cdots + \dfrac{101}{100^2}$

5 $a_1^2 + a_2^2 + a_3^2 + \cdots + a_n^2$

6 $a_1 + 2a_2 + 3a_3 + \cdots + na_n$

7 $\dfrac{1}{a_1} + \dfrac{2}{a_2} + \dfrac{3}{a_3} + \cdots + \dfrac{n}{a_n}$

Write down longhand versions of the following expressions:

8 $\sum_{r=1}^{5} r^3$

9 $\sum_{r=3}^{6} \dfrac{12}{r^2}$

10 $\sum_{r=1}^{20} 3 \times 2^r$

11 $\sum_{r=1}^{5} a_r$

12 $\sum_{r=4}^{8} \dfrac{r+1}{a_r}$

Use your calculator to evaluate

13 $\sum_{r=1}^{10} r^2$

14 $\sum_{r=1}^{100} \dfrac{20}{r^3}$

15 $\sum_{r=1}^{1000} \dfrac{1}{r}$

16 Use the facts that $\sum_{r=1}^{6} 2^r = 126$ and $\sum_{r=1}^{6} 3^r = 1092$ to deduce the values of

a) $\sum_{r=1}^{6} (2^r + 3^r)$

b) $\sum_{r=1}^{6} (5 \times 2^r)$

c) $\sum_{r=1}^{6} (3^r + 5)$

d) $\sum_{r=1}^{6} (2^{r+2})$

e) $\sum_{r=1}^{6} (3^r - 5 \times 2^r)$

Having studied this chapter you should know

- how to use definitions such as $u_n = n^2 + 1$ or $v_1 = 1$, $v_{n+1} = v_n + 2n + 1$ to calculate the first few terms of a sequence and deduce simple properties of the sequence
- how to use Σ notation as a shorthand for sums
- the three basic algebraic properties of Σ notation:

$$\sum_{r=1}^{n} (a_r + b_r) = \sum_{r=1}^{n} a_r + \sum_{r=1}^{n} b_r$$

$$\sum_{r=1}^{n} ca_r = c \sum_{r=1}^{n} a_r$$

$$\sum_{r=1}^{n} c = cn$$

REVISION EXERCISE

1 A sequence u_n is defined by $u_n = 8 + 100 \times 0.9^n$.
   a) Find the first four terms of the sequence.
   b) Describe what happens to the values of the sequence as $n \to \infty$.

2 Find the value of $\displaystyle\sum_{r=3}^{6} r^2$.

3 A sequence y_n is defined by $y_1 = 1000$, $y_{n+1} = 0.6y_n + 160$.
   a) Calculate the values of y_2, y_3 and y_4.
   b) Given that the sequence converges to a value L, determine the value of L.

4 A sequence of numbers x_n is generated by the rule

$$x_1 = 5, \, x_{n+1} = \frac{50}{x_n + 2}$$

   and the sequence converges to a limit M.
   a) Find, correct to three decimal places, the values of x_2, x_3 and x_4.
   b) Find the exact value of the constant M.

5 A sequence x_n is defined by $x_n = 2 \sin(60n°)$.
   a) Find the first eight values of the sequence.
   b) Find the value of x_{300}.
   c) Without using a calculator, find the value of $\displaystyle\sum_{r=1}^{300} x_r$.

6 A large insurance company initially has 2000 computers.
   Each year it expects to have to scrap 25% of its computers.
   Financial restrictions mean that it can only buy 300 new computers each year.
   If x_n denotes the number of computers that the company will have in n years' time:
   a) write down the value of x_0;
   b) show that $x_{n+1} = 0.75x_n + 300$;
   c) calculate the values of x_1, x_2 and x_3;
   d) if this pattern of computer disposal and purchase continues in the future, the number of computers owned by the company will tend to a limit L. Find the value of L.

7 **a)** Find integers p and q such that $\dfrac{8n+5}{2n-1} = p + \dfrac{q}{2n-1}$.

The nth term of a sequence of numbers u_n is defined by $\dfrac{8n+5}{2n-1}$:

b) find the first three terms of the sequence;

c) determine what happens to the terms of the sequence as $n \to \infty$;

d) calculate $\displaystyle\sum_{r=1}^{5} u_r$ giving your answer correct to two decimal places.

8 The sequence $u_1, u_2, u_3, \ldots$, where u_1 is a given real number, is defined for $n \geqslant 1$ by $u_{n+1} = 6 - u_n$.

i) If $u_1 = 2$

   a) Find the values of u_2, u_3, u_4 and u_5 and describe the behaviour of the sequence.

   b) Evaluate $\displaystyle\sum_{i=1}^{5} u_i$.

   c) Evaluate $\displaystyle\sum_{i=1}^{100} u_i$.

ii) Find the value of u_1 for which all values of the sequence are the same.

iii) If $u_1 = k$, prove that $\displaystyle\sum_{i=1}^{1000} u_i = 3000$.

9 Given that $\displaystyle\sum_{r=1}^{10} r^3 = 3025$ and $\displaystyle\sum_{r=1}^{10} 2^r = 2046$, find the values of

a) $\displaystyle\sum_{r=1}^{10} (r^3 + 2^r)$ **b)** $\displaystyle\sum_{r=1}^{10} (4r^3)$ **c)** $\displaystyle\sum_{r=1}^{10} (4 + r^3)$

5 Integration

The purpose of this chapter is to enable you to

- regard integration as the reverse process of differentiation
- evaluate indefinite and definite integrals
- calculate areas of regions bounded by a positive curve, the x axis and the lines $x = a$ and $x = b$
- calculate areas formed by negative curves; area between two curves; area between a curve and the y axis
- develop an initial understanding of the definite integral as the limit of a sum
- use the trapezium rule to estimate the values of definite integrals

Review of Differentiation

You have seen in the previous module how the gradient of a graph can be calculated from the equation of the graph. This chapter is concerned with trying to reverse the process and find the equation of a curve when given a full description of the gradient of the curve.

Since much of this chapter depends on a good understanding of the differentiation work encountered in C1, we start by briefly revising the key points:

- the gradient of a curve at a point is defined to be the gradient of the tangent to the curve at that point
- we write $\dfrac{dy}{dx}$ for the rule that gives the gradient of an (x, y) curve; we write $f'(x)$ for the gradient of the function $f(x)$
- if $y = x^n$ then $\dfrac{dy}{dx} = nx^{n-1}$
- if $y = a\,f(x) + b\,g(x)$, where a and b are constants, then $\dfrac{dy}{dx} = a\,f'(x) + b\,g'(x)$.

EXAMPLE 1

Find $\dfrac{dy}{dx}$ if

a) $y = 12x^3 + \dfrac{6}{x^2} + 5$

b) $y = \dfrac{x^2 - 5}{\sqrt{x}}$

EXAMPLE 1 (continued)

a)
$$y = 12x^3 + \frac{6}{x^2} + 5$$
$$\Rightarrow \quad y = 12x^3 + 6x^{-2} + 5$$
[differentiate] $\Rightarrow \quad \dfrac{dy}{dx} = 12 \times 3x^2 + 6 \times -2x^{-3} + 0$
$$\Rightarrow \quad \frac{dy}{dx} = 36x^2 - 12x^{-3}$$
$$\Rightarrow \quad \frac{dy}{dx} = 36x^2 - \frac{12}{x^3}$$

b)
$$y = \frac{x^2 - 5}{\sqrt{x}}$$
$$\Rightarrow \quad y = \frac{x^2}{\sqrt{x}} - \frac{5}{\sqrt{x}}$$
$$\Rightarrow \quad y = \frac{x^2}{x^{0.5}} - \frac{5}{x^{0.5}}$$
$$\Rightarrow \quad y = x^{1.5} - 5x^{-0.5}$$
[differentiate] $\Rightarrow \quad \dfrac{dy}{dx} = 1.5x^{0.5} - 5 \times -0.5x^{-1.5}$
$$\Rightarrow \quad \frac{dy}{dx} = 1.5x^{0.5} + 2.5x^{-1.5}$$

EXAMPLE 2

Find the gradient of the curve $y = (x^2 + 3)(2x - 3)$ at the point (2, 7).

$$y = (x^2 + 3)(2x - 3)$$
$$\Rightarrow \quad y = 2x^3 - 3x^2 + 6x - 9$$
$$\Rightarrow \frac{dy}{dx} = 6x^2 - 6x + 6$$

When $x = 2$, gradient $= 6 \times 2^2 - 6 \times 2 + 6 = 18$.

EXERCISE 1

Make sure you can do these questions before progressing to the next section. If necessary look back at the C1 work on differentiation to revise the topic thoroughly.

In questions 1–10, find $\dfrac{dy}{dx}$ if

1 $y = 2x^4 + 5x - 7$

2 $y = (3x^2 - 5)(2x^2 + 5)$

3 $y = \dfrac{4}{x^2}$

4 $y = \dfrac{4}{x} - \dfrac{3}{x^3}$

5 $y = 3x^2 + 5 - \dfrac{4}{x^3}$

6 $y = 6\sqrt{x} - \dfrac{4}{\sqrt{x}}$

7 $y = \dfrac{3x^3 + 5}{x^2}$

8 $y = (x^2\sqrt{x})^3$

9 $y = (\sqrt{x} - 3)^2$

10 $y = 6\sqrt[3]{x}$

In questions 11–15, find the gradient of the curve at the given point:

11 $y = 2x^3 - 7x^2 + 5x + 4$ at $(1, 4)$

12 $y = \dfrac{12}{x}$ at $(2, 6)$

13 $y = 8\sqrt{x}$ at $(4, 16)$

14 $y = \left(x^2 - \dfrac{4}{x}\right)^2$ at $(2, 4)$

15 $y = \dfrac{36}{x^2} - \dfrac{9}{x}$ at $(-3, 7)$

Integration as the Reverse of Differentiation

Suppose a curve is such that the gradient at any point (x, y) on the curve is $3x^2 + 5$.

This means that $\dfrac{dy}{dx} = 3x^2 + 5$.

What can be said about y?

You know $3x^2$ is the derivative of x^3 and that 5 is the derivative of $5x$. It is therefore possible that $y = x^3 + 5x$.

However, since the derivative of any constant number is 0 it is equally possible that

$$y = x^3 + 5x + 7 \quad \text{or} \quad y = x^3 + 5x - 82.$$

The equation $y = x^3 + 5x + c$, where c is a constant, is the **general solution** of the equation $\dfrac{dy}{dx} = 3x^3 + 5$.

EXAMPLE 3

Find y if $\dfrac{dy}{dx} = 8x^3 + 6x^2 - 5$.

You know $8x^3$ is the derivative of $2x^4$.
You know $6x^2$ is the derivative of $2x^3$.
You know 5 is the derivative of $5x$.

Remember to include
$+ c$ in your answer.

So $\dfrac{dy}{dx} = 8x^3 + 6x^2 - 5 \implies y = 2x^4 + 2x^3 - 5x + c$.

EXAMPLE 4

Find the curve that passes through the point $(1, 5)$ and has gradient $9x^2 + \dfrac{8}{x^2}$.

You have $\dfrac{dy}{dx} = 9x^2 + \dfrac{8}{x^2}$.

You know $9x^2$ is the derivative of $3x^3$.

You know $\dfrac{8}{x^2}$ is the derivative of $-\dfrac{8}{x}$.

The thought process leading to this might be the derivative of $\dfrac{1}{x}$ is $-\dfrac{1}{x^2}$ so the derivative of $\dfrac{k}{x}$ is $-\dfrac{k}{x^2}$. In particular, if $k = -8$, then the derivative of $-\dfrac{8}{x}$ is $\dfrac{8}{x^2}$.

EXAMPLE 4 (continued)

The general solution of

$$\frac{dy}{dx} = 9x^2 + \frac{8}{x^2}$$

is therefore $y = 3x^3 - \frac{8}{x} + c$.

The value of c can be found by using the fact that the curve must pass through the point (1, 5). Substituting $x = 1$ and $y = 5$ into the general solution gives

$$5 = 3 - 8 + c \quad \Rightarrow \quad c = 10.$$

The required solution is therefore

$$y = 3x^3 - \frac{8}{x} + 10.$$

The Notation of Integration

You have seen that if $\dfrac{dy}{dx} = 3x^2 + 5$ then $y = x^3 + 5x + c$.

$x^3 + 5x + c$ is called the **indefinite integral** of $3x^2 + 5$; this is written as

$$\int (3x^2 + 5) \, dx = x^3 + 5x + c$$

and you would say "the integral of $3x^2 + 5$ with respect to x is $x^3 + 5x + c$".

You have also seen that if $\dfrac{dy}{dx} = 8x^3 + 6x^2 - 5$ then $y = 2x^4 + 2x^3 - 5x + c$ so you can write

$$\int 8x^3 + 6x^2 - 5 \, dx = 2x^4 + 2x^3 - 5x + c.$$

Similarly, you have seen that if $\dfrac{dy}{dx} = 9x^2 + \dfrac{8}{x^2}$ then $y = 3x^3 - \dfrac{8}{x} + c$ so you can write

$$\int \left(9x^2 + \frac{8}{x^2}\right) dx = 3x^3 - \frac{8}{x} + c.$$

EXAMPLE 5

Find $\displaystyle\int (4x^3 - 9x^2 + 5) \, dx$.

$$\int (4x^3 - 9x^2 + 5) \, dx = x^4 - 3x^3 + 5x + c.$$

This is really the same problem as

"find y if $\dfrac{dy}{dx} = 4x^3 - 9x^2 + 5$".

You know that

$4x^3$ is the derivative of x^4;
$9x^2$ is the derivative of $3x^3$;
5 is the derivative of $5x$

so $\dfrac{dy}{dx} = 4x^3 - 9x^2 + 5 \Rightarrow y = x^4 - 3x^3 + 5x + c$.

EXERCISE 2

Find y if

1 $\dfrac{dy}{dx} = 4x + 3$

2 $\dfrac{dy}{dx} = 3x^2 + 6x - 2$

3 $\dfrac{dy}{dx} = \dfrac{-4}{x^2}$

4 $\dfrac{dy}{dx} = 8x^3 - 6x$

5 $\dfrac{dy}{dx} = x^2 + 3x + 1$

6 $\dfrac{dy}{dx} = 10x^4 - 12x^3 + 8x - 3$

7 $\dfrac{dy}{dx} = \dfrac{10}{x^2}$

8 $\dfrac{dy}{dx} = \dfrac{10}{x^3}$

9 Find the equation of the curve that passes through the point (1, 8) and has gradient $6x - 11$.

10 Find the equation of the curve that passes through the point (−1, 2) and has gradient $12x^2 - 8x + 1$.

Find the following integrals:

11 $\displaystyle\int (9x^2 - 8x + 5)\, dx$

12 $\displaystyle\int (3x + 2)\, dx$

13 $\displaystyle\int (7x - 3)\, dx$

14 $\displaystyle\int \dfrac{9}{x^4}\, dx$

15 $\displaystyle\int \left(6x^2 - \dfrac{12}{x^3}\right) dx$

The Integral of x^n

Using your basic rule for differentiation, you can obtain a rule for integration.

You know that the derivative of x^{n+1} is $(n + 1)x^n$.

The derivative of $\dfrac{1}{n+1} x^{n+1}$ is therefore $\dfrac{1}{n+1}(n + 1)x^n$ which simplifies to x^n.

Since integration is the reverse of differentiation you can therefore rewrite this as

$$\int x^n\, dx = \frac{1}{n+1} x^{n+1} + c.$$

You can't calculate $\dfrac{1}{n+1}$ when $n = -1$ so this result for integration cannot hold when $n = -1$.

You will see how to find $\displaystyle\int x^{-1}\, dx$ in module C3.

You therefore have the very important result

$$\int x^n\, dx = \frac{x^{n+1}}{n+1} + c \qquad (n \neq -1)$$

As special cases of this result you have:

$$\int x^3 \, dx = \frac{x^4}{4} + c = \frac{1}{4} x^4 + c$$

$$\int x^2 \, dx = \frac{x^3}{3} + c = \frac{1}{3} x^3 + c$$

$$\int x \, dx = \int x^1 \, dx = \frac{x^2}{2} + c = \frac{1}{2} x^2 + c$$

$$\int 1 \, dx = \int x^0 \, dx = \frac{x^1}{1} + c = x + c$$

$$\int \frac{1}{x^2} \, dx = \int x^{-2} \, dx = \frac{x^{-1}}{-1} + c = -\frac{1}{x} + c$$

$$\int \frac{1}{x^3} \, dx = \int x^{-3} \, dx = \frac{x^{-2}}{-2} + c = -\frac{1}{2x^2} + c$$

and

$$\int \sqrt{x} \, dx = \int x^{\frac{1}{2}} \, dx = \frac{x^{\frac{3}{2}}}{\frac{3}{2}} + c = \frac{2}{3} x^{\frac{3}{2}} + c.$$

> Remember that dividing by $\frac{3}{2}$ is the same as multiplying by $\frac{2}{3}$.

Although the variable in an integration is not always x, the general rule still applies. For example,

$$\int t^2 \, dt = \frac{1}{3} t^3 + c$$

and

$$\int z^n \, dz = \frac{z^{n+1}}{n+1} + c \qquad \text{(provided } n \neq -1\text{)}.$$

A modest amount of algebraic manipulation will often enable you to rewrite a complicated function into an equivalent form that can be integrated.

EXAMPLE 6

Find

a) $\int \frac{20}{x^5} \, dx$ **b)** $\int \frac{8}{\sqrt[3]{x}} \, dx$ **c)** $\int (\sqrt{t})^5 \, dt$

EXAMPLE 6 (continued)

a) $\int \dfrac{20}{x^5}\,dx = \int 20 \times \dfrac{1}{x^5}\,dx$

$\quad\quad\quad = \int 20x^{-5}\,dx$

$\quad\quad\quad = 20\dfrac{x^{-4}}{-4} + c$

$\quad\quad\quad = -5x^{-4} + c = -\dfrac{5}{x^4} + c$

b) $\int \dfrac{8}{\sqrt[3]{x}}\,dx = \int 8x^{-\frac{1}{3}}\,dx$ — Since $\dfrac{8}{\sqrt[3]{x}} = \dfrac{8}{x^{\frac{1}{3}}} = 8x^{-\frac{1}{3}}.$

$\quad\quad\quad = 8\dfrac{x^{\frac{2}{3}}}{\dfrac{2}{3}} + c$ — Dividing by $\dfrac{2}{3}$ is the same as multiplying by $\dfrac{3}{2}$.

$\quad\quad\quad = 8 \times \dfrac{3}{2}x^{\frac{2}{3}} + c$

$\quad\quad\quad = 12x^{\frac{2}{3}} + c$

c) $\int (\sqrt{t})^5\,dt = \int t^{\frac{5}{2}}\,dt$ — Since $(\sqrt{t})^5 = (t^{\frac{1}{2}})^5 = t^{\frac{5}{2}}.$

$\quad\quad\quad = \dfrac{t^{\frac{7}{2}}}{\dfrac{7}{2}} + c$ — Dividing by $\dfrac{7}{2}$ is the same as multiplying by $\dfrac{2}{7}$.

$\quad\quad\quad = \dfrac{2}{7}t^{\frac{7}{2}} + c$

EXAMPLE 7

Find $\int (2x + 3)^2\,dx$.

Expanding $(2x + 3)^2$ gives

$\int (2x + 3)^2\,dx = \int (4x^2 + 12x + 9)\,dx$

$\quad\quad\quad = 4\dfrac{x^3}{3} + 12\dfrac{x^2}{2} + 9x + c$

$\quad\quad\quad = \dfrac{4}{3}x^3 + 6x^2 + 9x + c.$

EXAMPLE 8

Find $\displaystyle\int \frac{(2+x^2)^2}{x^2}\, dx$.

$\displaystyle\int \frac{(2+x^2)^2}{x^2}\, dx = \int \frac{4+4x^2+x^4}{x^2}\, dx = \int \frac{4}{x^2}+4+x^2\, dx = \int 4x^{-2}+4+x^2\, dx$

$\displaystyle = 4\frac{x^{-1}}{-1}+4x+\frac{x^3}{3}+c = \frac{-4}{x}+4x+\frac{1}{3}x^3+c.$

EXAMPLE 9

Find $\displaystyle\int \frac{1+p^2}{\sqrt{p}}\, dp$.

$\displaystyle\int \frac{1+p^2}{\sqrt{p}}\, dp = \int \frac{1}{\sqrt{p}}+\frac{p^2}{\sqrt{p}}\, dp$ — The first task is to break the function up into a sum or difference of powers of p.

$\displaystyle = \int p^{-\frac{1}{2}}+p^{\frac{3}{2}}\, dx$ — Now use the integration rule $\displaystyle\int p^n\, dp = \frac{1}{n+1}p^{n+1}.$

$\displaystyle = \frac{p^{\frac{1}{2}}}{\frac{1}{2}}+\frac{p^{\frac{5}{2}}}{\frac{5}{2}}+c$

$\displaystyle = 2p^{\frac{1}{2}}+\frac{2}{5}p^{\frac{5}{2}}+c.$ — Dividing by $\frac{1}{2}$ is the same as multiplying by 2; dividing by $\frac{5}{2}$ is the same as multiplying by $\frac{2}{5}$.

EXERCISE 3

Find

1 $\displaystyle\int 5x^2\, dx$

2 $\displaystyle\int 8u^3\, du$

3 $\displaystyle\int 6\sqrt{s}\, ds$

4 $\displaystyle\int \frac{12}{x^2}\, dx$

5 $\displaystyle\int \frac{1}{\sqrt{x}}\, dx$

6 $\displaystyle\int {}^5\sqrt{x}\, dx$

7 $\displaystyle\int t^{\frac{2}{5}}\, dt$

8 $\displaystyle\int \frac{4}{{}^3\sqrt{y}}\, dy$

9 $\displaystyle\int \frac{18}{t^4}\, dt$

10 $\displaystyle\int 10x^{\frac{2}{3}}\, dx$

11 $\displaystyle\int 5x^{-\frac{3}{4}}\, dx$

12 $\displaystyle\int \sqrt{x^3}\, dx$

13 $\displaystyle\int (3x-2)(5x+4)\, dx$

14 $\displaystyle\int (2x+1)^2\, dx$

15 $\displaystyle\int \frac{3+x^3}{x^2}\, dx$

16 $\displaystyle\int \frac{4+u}{u^3}\, du$

17 $\displaystyle\int (x^2-2)(x^3+1)\, dx$

18 $\displaystyle\int \frac{3t+t^5}{t^3}\, dt$

19 $\displaystyle\int \frac{1+x}{\sqrt{x}}\, dx$

20 $\displaystyle\int \frac{x^4+5}{x^2}\, dx$

21 If $\dfrac{dx}{dt} = 8t + 5$ and $x = 2$ when $t = 0$, find an expression for x in terms of t. Hence find the value of x when $t = 2$.

22 If $\dfrac{dz}{dp} = 12\sqrt{p}$ and $z = 10$ when $p = 4$, find an expression for z in terms of p.

23 If $\dfrac{dx}{dt} = 12 - 9.8t$ and $x = 2$ when $t = 0$, find the value of x when $t = 2$.

24 If $\dfrac{dy}{dt} = (5t - 1)(3t - 5)$ and $y = 4$ when $t = 1$, find an expression for y in terms of t.

⑪ Estimating the Area under a Curve

Consider the area under the graph of $y = x^2$ between $x = 0$ and $x = 3$.

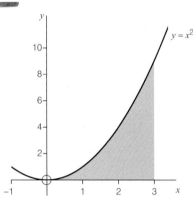

The area can be approximated by a series of rectangles of width 1 unit as shown in the diagram:

$$\text{area} \approx 0 \times 1 + 1 \times 1 + 4 \times 1 = 5.$$

Clearly this is a very poor estimate since large parts of the required area have been omitted.

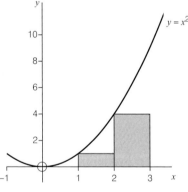

A better estimate can be obtained by using smaller rectangles of width 0.5 unit.

The improved estimate is

$$\begin{aligned}\text{area} \approx\ & 0^2 \times 0.5 + 0.5^2 \times 0.5 + 1^2 \times 0.5 \\ & + 1.5^2 \times 0.5 + 2^2 \times 0.5 + 2.5^2 \times 0.5 \\ =\ & 6.875.\end{aligned}$$

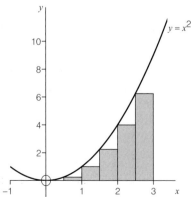

Obviously you can continue this process and obtain better and better approximations to the required area by considering many more, much thinner rectangular strips. However the arithmetic becomes increasingly cumbersome so progress will require a computer.

In general, suppose the area under $y = f(x)$ between $x = a$ and $x = b$ is split into n rectangles each of width δx.

Then the area of the ith rectangle is

$$f(x_i)\delta x$$

so the estimate of the area under the curve will be:

$$\text{area} \approx \sum_{i=1}^{n} f(x_i)\delta x.$$

We can express the fact that the accuracy improves as we take more and more thinner rectangles by saying that:

$$\text{area} = \lim_{\delta x \to 0} \sum_{i=1}^{n} f(x_i)\delta x.$$

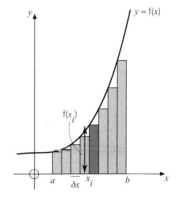

The Excel worksheet shown in the screen dump below allows us to split the area under the graph of $y = kx^n$ between $x = a$ and $x = b$ into a large number (up to 10 000) of rectangular strips and estimates the area under the curve by calculating the sum of the areas of the rectangular strips.

	A	B	C	D	E	F	G	H	I
1	k =	1							
2	n =	2							
3	a =	0							
4	b =	3							
5	No. of strips =	3000							
6									
7									
8									
9	strip width =	0.001		Area estimate =	8.9955005				
10									
11									
12	strip	x	y	rectangle area					
13	1	0	0	0					
14	2	0.001	0.000001	0.000000001					
15	3	0.002	0.000004	0.000000004					
16	4	0.003	0.000009	0.000000009					
17	5	0.004	0.000016	0.000000016					
18	6	0.005	0.000025	0.000000025					
19	7	0.006	0.000036	0.000000036					
20	8	0.007	0.000049	0.000000049					
21	9	0.008	0.000064	0.000000064					
22	10	0.009	0.000081	0.000000081					
23	11	0.01	0.0001	0.0000001					
24	12	0.011	0.000121	0.000000121					
25	13	0.012	0.000144	0.000000144					
26	14	0.013	0.000169	0.000000169					
27	15	0.014	0.000196	0.000000196					
28	16	0.015	0.000225	0.000000225					

Sheet1 / Sheet2 / Sheet3

Formulae

cell B9   =(B4-B3)/B5
cell E9   =SUM(D13:D10012)
cell A13 =1
cell B13 =B3
cell C13 =IF(B13="","",B1*B13^B2)
cell D13 =IF(B13="","",C13*B9)
cell A14 =IF(A13<B5,A13+1,"")
cell B14 =IF(A14="","",B13+B9)
cell C14 =IF(B14="","",B1*B14^B2)
cell D14 =IF(B14="","",C14*B9)
row 14 should then be copied down to
row 10012.

This printout shows that, on the basis of 3000 rectangular strips, the estimate of the area under $y = x^2$ between $x = 0$ and $x = 3$ is 8.996 to three decimal places.

Using the Excel worksheet, these estimates of the area under $y = x^2$ between $x = 0$ and $x = 3$ can be rapidly obtained:

Number of rectangular strips used	Estimate of the area under $y = x^2$ between $x = 0$ and $x = 3$
30	8.555
60	8.766
300	8.955
600	8.978
3000	8.996
6000	8.998

and it certainly appears that the estimates are getting closer and closer to 9.

EXERCISE 4

Make sure you work through this exercise, which uses the spreadsheet to investigate areas under curves and motivate the results of the next section.

1 Use this spreadsheet to obtain good estimates of the area under the curve $y = x^2$
   a) between $x = 0$ and $x = 1$;
   b) between $x = 0$ and $x = 2$;
   c) between $x = 0$ and $x = 4$;
   d) between $x = 0$ and $x = 5$.

Find a rule which gives the area under the curve $y = x^2$ between $x = 0$ and $x = b$.

2 Use this spreadsheet to obtain good estimates of the area under the curve $y = x^2$
   a) between $x = 1$ and $x = 2$;
   b) between $x = 1$ and $x = 3$;
   c) between $x = 1$ and $x = 4$;
   d) between $x = 1$ and $x = 5$.

Find a rule which gives the area under the curve $y = x^2$ between $x = 1$ and $x = b$.
Try to explain the connection between the rules you have obtained in question 1 and question 2.

3 Use this spreadsheet to obtain good estimates of the area under the curve $y = x^2$
   a) between $x = 2$ and $x = 3$;
   b) between $x = 2$ and $x = 4$;
   c) between $x = 2$ and $x = 5$.

Find a rule which gives the area under the curve $y = x^2$ between $x = 2$ and $x = b$.
Try to explain the connection between the rules you have obtained in question 1 and question 3.

4 Find a rule which gives the area under the curve $y = x^2$ between $x = a$ and $x = b$, where $0 < a < b$.

5 Use this spreadsheet to obtain good estimates of the area under the curve $y = x^3$
   a) between $x = 0$ and $x = 1$;
   b) between $x = 0$ and $x = 2$;
   c) between $x = 0$ and $x = 3$;
   d) between $x = 0$ and $x = 4$.

Find a rule which gives the area under the curve $y = x^3$ between $x = 0$ and $x = b$.
Write down a rule for the area under the curve $y = x^3$ between $x = a$ and $x = b$, where $0 < a < b$.

6 Find a rule for the area under the curve $y = x^4$ between $x = a$ and $x = b$, where $0 < a < b$.

7 If $0 < a < b$, prove that the area under the graph of $y = x$ between $x = a$ and $x = b$ is given by $\dfrac{1}{2}b^2 - \dfrac{1}{2}a^2$.

Use your previous results to complete the following table:

Curve	Area under curve between $x = a$ and $x = b$
$y = x$	$\dfrac{1}{2}b^2 - \dfrac{1}{2}a^2$
$y = x^2$	
$y = x^3$	
$y = x^4$	

What connection have these results with the idea of integration?

Calculating the Area of the Region Formed by a Positive Curve, the x Axis and Two Lines Parallel to the y Axis

You have seen in the previous exercise that if f(x) is positive then the area under the curve y = f(x) between $x = a$ and $x = b$ appears to be the same as

(the value of the integral of f(x) when $x - b$) – (the value of the integral of f(x) when $x - a$).

A justification of this important result appears in the Extension section at the end of this chapter.

The usual way of performing this calculation is by means of a **definite integral** — that is an integral with limits. Example 10 below shows how a definite integral is evaluated.

EXAMPLE 10

> It is a good idea to produce a quick sketch of the region whose area is being calculated. Sometimes you may need to use a graphical calculator to do this quickly.

Calculate the area of the closed region bounded by the curve $y = x^2$, the x axis and the lines $x = 1$ and $x = 4$.

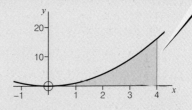

The area will be given by

(the value of the integral of x^2 when $x = 4$) – (the value of the integral of x^2 when $x = 1$).

This calculation is usually presented in the following way.

$$\text{Area} = \int_1^4 x^2 \, \mathrm{d}x$$

> This reads as the **definite integral** of x^2 between $x = 1$ and $x = 4$.
> 1 is called the **lower limit** of the integral;
> 4 is called the **upper limit** of the integral.

$$\Rightarrow \quad \text{Area} = \left[\frac{1}{3} x^3 \right]_1^4$$

> The first step is evaluating a definite integral is to do the integration.
> The integral of x^2 is $\frac{1}{3} x^3$.

$$\Rightarrow \quad \text{Area} = \left[\frac{1}{3} x^3 \right]_1^4 = \left(\frac{1}{3} 4^3 \right) - \left(\frac{1}{3} 1^3 \right)$$

> This is written inside square brackets with the limits appearing immediately after the brackets.

$$\Rightarrow \quad \text{Area} = \left[\frac{1}{3} x^3 \right]_1^4 = \left(\frac{1}{3} 4^3 \right) - \left(\frac{1}{3} 1^3 \right) = \frac{64}{3} - \frac{1}{3} = \frac{63}{3} = 21.$$

> The second step is to find the value of the integral when x takes its upper value and then subtract the value of the integral when x takes its lower value.

Notice that the " $+ c$" was omitted in the definite integral. If it had been included the working would be:

$$\text{area} = \left[\frac{1}{3} x^3 + c \right]_1^4 = \left(\frac{1}{3} 4^3 + c \right) - \left(\frac{1}{3} 1^3 + c \right) = \frac{64}{3} + c - \frac{1}{3} - c = \frac{63}{3} = 21$$

and it can be seen that the c terms just cancel each other out.

When evaluating a definite integral the " $+ c$" can safely be omitted.

EXAMPLE 11

Calculate the area under the curve $y = 8x^3 + 6x + 2$ between $x = 0$ and $x = 2$.

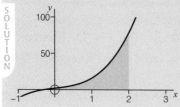

$$\text{Area} = \int_0^2 8x^3 + 6x + 2 \; dx$$

$$= [2x^4 + 3x^2 + 2x]_0^2$$

$$= (32 + 12 + 4) - (0 + 0 + 0)$$

$$= 48.$$

N.B. Your graphical calculator may well be able to evaluate definite integrals: you may use it to check answers but you should always show full details of your working.

To summarise: you now have two formulations for the area under a positive function, $y = f(x)$, between $x = a$ and $x = b$. It can be written as the limit of a sum of rectangular strips or it can be written as a definite integral.

If the curve $y = f(x)$ is positive between $x = a$ and $x = b$ then the area of the closed region bounded by the curve, the x axis and the lines $x = a$ and $x = b$ is given by

$$\text{Area} = \lim_{\delta x \to 0} \sum_{i=1}^{n} f(x_i)\delta x = \int_a^b f(x) \; dx = \int_a^b y \; dx$$

In later modules, this equivalence of a definite integral with the limit of a sum will be very useful in establishing further applications of integration.

For the moment though the important result is

$$\textbf{Area} = \int_a^b \textbf{f}(x) \, dx = \int_a^b y \, dx.$$

EXAMPLE 12

Evaluate

a) $\displaystyle\int_3^{10} \frac{18}{x^3} \, dx$

b) $\displaystyle\int_3^{100} \frac{18}{x^3} \, dx$

c) $\displaystyle\int_3^{1000} \frac{18}{x^3} \, dx$ and interpret your answers as areas.

a) $\displaystyle\int_3^{10} \frac{18}{x^3} \, dx = \left[-\frac{9}{x^2} \right]_3^{10} = (-0.09) - (-1) = 0.91$

This is the area of the closed region bounded by the curve $y = \dfrac{18}{x^3}$, the x axis and the lines $x = 3$ and $x = 10$.

b) $\displaystyle\int_3^{100} \frac{18}{x^3} \, dx = \left[-\frac{9}{x^2} \right]_3^{100} = (-0.0009) - (-1) = 0.9991$

This is the area of the closed region bounded by the curve $y = \dfrac{18}{x^3}$, the x axis and the lines $x = 3$ and $x = 100$.

c) $\displaystyle\int_3^{1000} \frac{18}{x^3} \, dx = \left[-\frac{9}{x^2} \right]_3^{1000} = (-0.000009) - (-1) = 0.999991$

This is the area of the closed region bounded by the curve $y = \dfrac{18}{x^3}$, the x axis and the lines $x = 3$ and $x = 1000$.

To generalise these three results you can write

$$\int_3^N \frac{18}{x^3}\, dx = \left[-\frac{9}{x^2} \right]_3^N = \left(-\frac{9}{N^2} \right) - (-1) = 1 - \frac{9}{N^2}$$

and deduce that if A_N denotes the area of the closed region bounded by the curve $y = \dfrac{18}{x^3}$, the x axis and the lines $x = 3$ and $x = N$ then $A_N = 1 - \dfrac{9}{N^2}$.

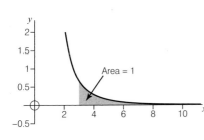

Area = 1

As $N \rightarrow \infty$, $A_N \rightarrow 1$ so you can say that 1 is the area of the region to the right of $x = 3$ bounded by the curve $y = \dfrac{18}{x^3}$ and the x axis.

The calculation of this area can be simplified by writing

$$\text{area} = \int_3^\infty \frac{18}{x^3}\, dx = \left[-\frac{9}{x^2} \right]_3^\infty = \left(-\frac{9}{\infty^2} \right) - (-1) = 0 - (-1) = 1.$$

You could determine the value of $\dfrac{9}{\infty^2}$ by first thinking that "9 divided by a very large number squared gives an answer very close to 0" and then deducing that $\dfrac{9}{\infty^2} = 0$.

EXAMPLE 13

Evaluate $\displaystyle\int_2^\infty \frac{24}{x^5}\, dx$ and interpret your answer as an area.

$$\int_2^\infty \frac{24}{x^5}\, dx = \left[-\frac{6}{x^4} \right]_2^\infty$$

$$= \left(\frac{-6}{\infty^4} \right) - \left(-\frac{6}{16} \right)$$

$$= 0 + \frac{3}{8}$$

$$= \frac{3}{8}.$$

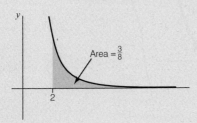

Area = $\frac{3}{8}$

This integral gives the area of the region between the curve $y = \dfrac{24}{x^5}$ and the x axis to the right of $x = 2$.

EXERCISE 5

In questions 1–9, evaluate the following definite integrals and describe carefully the region whose area has been found:

1 $\displaystyle\int_1^2 \frac{1}{x^2}\,dx$

2 $\displaystyle\int_1^2 (x+1)(x+2)\,dx$

3 $\displaystyle\int_0^4 \sqrt{x}\,dx$

4 $\displaystyle\int_1^2 x^2 - x\,dx$

5 $\displaystyle\int_1^3 \frac{1+x^3}{x^2}\,dx$

6 $\displaystyle\int_0^4 v\sqrt{v}\,dv$

7 $\displaystyle\int_1^\infty \frac{4}{x^3}\,dx$

8 $\displaystyle\int_0^9 \frac{1}{\sqrt{t}}\,dt$

9 $\displaystyle\int_2^\infty \frac{6+x}{x^3}\,dx$

Find the area of the shaded regions in each of the following diagrams:

10
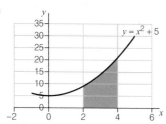

11
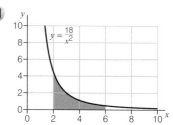

12
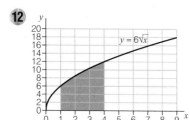

In questions 13–17, find the area of the closed region bounded by the given curve, the x axis and the two given lines parallel to the y axis. Illustrate your answers with a sketch graph.

13 $y = x + 3x^2$ between $x = 1$ and $x = 2$.

14 $y = \sqrt{x}$ between $x = 4$ and $x = 25$.

15 $y = \dfrac{1}{x^2}$ between $x = 1$ and $x = 4$.

16 $y = \dfrac{72}{x^4}$ between $x = 2$ and $x = \infty$.

17 $y = (x^2 + 3)^2$ between $x = 0$ and $x = 1$.

18 Find the area of the region between the curve $y = \dfrac{100}{x^2}$ and the x axis to the right of $x = 4$.

19 Find the area of the region between the curve $y = \dfrac{1}{x\sqrt{x}}$ and the x axis to the right of $x = 1$.

20 Find the value of p if $\displaystyle\int_1^\infty \frac{2}{x^2}\,dx = 3\int_1^p \frac{2}{x^2}\,dx$.

Further Area Calculations

Areas of Regions that are Below the x Axis

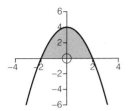

The diagram shows the graph of $y = 4 - x^2$.

The area of the closed region formed by the curve and the x axis is

$$\int_{-2}^{2} 4 - x^2 \, dx = \left[4x - \frac{1}{3}x^3 \right]_{-2}^{2} = \frac{16}{3} - \left(-\frac{16}{3} \right) = \frac{32}{3}.$$

The diagram now shows the graph of $y = x^2 - 4$.

Since this is a reflection in the x axis of the previous graph, the shaded area must also be $\dfrac{32}{3}$.

However, if you evaluate the integral $\displaystyle\int_{-2}^{2} x^2 - 4 \, dx$ you will obtain a negative answer:

$$\int_{-2}^{2} x^2 - 4 \, dx = \left[\frac{1}{3}x^3 - 4x \right]_{-2}^{2} = \left(-\frac{16}{3} \right) - \left(\frac{16}{3} \right) = -\frac{32}{3}.$$

So, for areas under the x axis the integral gives a negative number: the minus sign must be ignored if you wish to evaluate the area.

EXAMPLE 14

Sketch the graph of $y = 2x^2 + 4x - 6$ and calculate the area of the closed region formed by the curve and the x axis.

$y = 2x^2 + 4x - 6 = 2(x + 3)(x - 1)$

so the graph is $\cup$ shaped and passes through $(-3, 0)$ and $(1, 0)$.

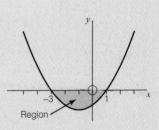

The region lies beneath the x axis so the integral will give a negative answer.

$$\int_{-3}^{1} 2x^2 + 4x - 6 \, dx = \left[\frac{2}{3}x^3 + 2x^2 - 6x \right]_{-3}^{1}$$

$$= \left(\frac{2}{3} + 2 - 6 \right) - \left((-18) + 18 - (-18) \right)$$

$$= \left(-\frac{10}{3} \right) - 18$$

$$= -\frac{64}{3}$$

so the area of the region is $\dfrac{64}{3}$ units^2.

If $f(x)$ is negative between $x = a$ and $x = b$ then $\displaystyle\int_a^b f(x)\,dx$ will give a negative answer and the area of the region formed by the curve $y = f(x)$, the x axis and the lines $x = a$ and $x = b$ is given by $-\displaystyle\int_a^b f(x)\,dx$.

If the region whose area is to be calculated lies on both sides of the x axis then the area above the x axis and the area below the x axis must **be calculated separately** and the two numerical results added together.

EXAMPLE 15

Find the area of the closed region formed by the curve $y = x^2 - 3x$, the x axis and the lines $x = 1$ and $x = 4$.

Again, a sketch graph showing the region should be drawn:

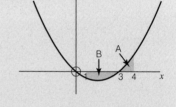

$$y = x^2 - 3x = x(x - 3).$$

The curve will be $\cup$ shaped and passes through $(0, 0)$ and $(3, 0)$.

The curve crosses the x axis at $x = 3$ so you should consider the area between 1 and 3 and the area between 3 and 4 **separately**.

Between $x = 3$ and $x = 4$ the function is positive so

$$\text{area A} = \int_3^4 x^2 - 3x\,dx = \left[\frac{1}{3}x^3 - \frac{3}{2}x^2\right]_3^4 = \left(\frac{8}{3}\right) - \left(-\frac{9}{2}\right) = \frac{11}{6}.$$

Between $x = 1$ and $x = 3$ the function is negative.

$$\int_1^3 x^2 - 3x\,dx = \left[\frac{1}{3}x^3 - \frac{3}{2}x^2\right]_1^3 = \left(-\frac{9}{2}\right) - \left(-\frac{7}{6}\right) = -\frac{10}{3}$$

so

$$\text{area B} = \frac{10}{3}.$$

$$\text{Total area} = \text{area A} + \text{B} = \frac{10}{3} + \frac{11}{6} = \frac{31}{6}.$$

Area Between Two Curves

EXAMPLE 16

Find the area between the curves $y = 4x^2$ and $y = x^3$.

The diagram shows a sketch of the two curves:

You must first find the points of intersection of the two curves:

$$\left. \begin{array}{r} y = 4x^2 \\ y = x^3 \end{array} \right\} \Rightarrow x^3 = 4x^2$$

$$\Rightarrow \quad x^3 - 4x^2 = 0$$
$$\Rightarrow \quad x^2(x - 4) = 0$$
$$\Rightarrow \quad x = 0 \text{ or } 4$$
$$\Rightarrow \quad A(0, 0) \text{ and } B(4, 64).$$

Area between curves = area under $y = 4x^2$ between 0 and 4
$$ - \text{area under } y = x^3 \text{ between 0 and 4}$$

$$= \int_0^4 4x^2 \, dx - \int_0^4 x^3 \, dx$$

$$= \int_0^4 4x^2 - x^3 \, dx = \left[\frac{4}{3}x^3 - \frac{1}{4}x^4 \right]_0^4 = \frac{64}{3}.$$

In general, the area between the curves $y = f(x)$ and $y = g(x)$ and the lines $x = a$ and $x = b$ is

$$\int_a^b f(x) - g(x) \, dx$$

provided $f(x) \geqslant g(x)$ for all values of x between a and b.

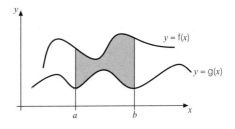

Note that the two functions **do not** need to be positive. You simply require that $f(x) \geqslant g(x)$ for all values of x between a and b.

You might think of this result as

$$\text{area between two curves} = \int_a^b (y_{top} - y_{bottom}) \, dx.$$

A **formal justification** of this result can be found in the Extension section at the end of the chapter.

EXAMPLE 17

Find the area of the closed region between the curve $y = x^2 - 4$ and the line $y = 3x$.

To find the points A and B you must solve the simultaneous equations

$$\left. \begin{array}{l} y = x^2 - 4 \\ y = 3x \end{array} \right\}$$

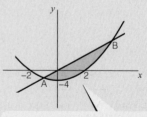

$\Rightarrow \quad x^2 - 4 = 3x$

$\Rightarrow \quad x^2 - 3x - 4 = 0$

$\Rightarrow \quad (x - 4)(x + 1) = 0$

$\Rightarrow \quad x = 4 \text{ or } -1$

$\Rightarrow \quad A(-1, -3) \text{ and } B(4, 12).$

Remember the sketch diagram!

$$\text{Area} = \int_a^b (y_{top} - y_{bottom}) \, dx$$

$$= \int_{-1}^4 ((3x) - (x^2 - 4)) \, dx$$

$$= \int_{-1}^4 (3x + 4 - x^2) \, dx = \left[\frac{3}{2} x^2 + 4x - \frac{1}{3} x^3 \right]_{-1}^4$$

$$= \left(24 + 16 - \frac{64}{3} \right) - \left(\frac{3}{2} - 4 + \frac{1}{3} \right) = \frac{56}{3} - \left(-\frac{13}{6} \right) = \frac{125}{6}.$$

EXAMPLE 18

The diagram shows the graph of $y = x^3$ and the tangent to the graph at the point P(4, 64).

a) Find the equation of the tangent.

b) Find the point Q where the curve and the tangent meet again.

c) Find the area of the closed region bounded by the curve and the tangent.

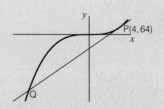

a) $y = x^3 \Rightarrow \dfrac{dy}{dx} = 3x^2 \Rightarrow$ gradient at P(4, 64) $= 3 \times 4^2 = 48.$

The tangent is the line through (4, 64) with gradient 48 and has equation

$$y - 64 = 48(x - 4)$$

$$\Rightarrow \quad y = 48x - 128.$$

Look back at the C1 chapter "Applications of Differentiation" if you have forgotten the work on tangents and normals.

EXAMPLE 18 (continued)

b) To find Q you must solve the simultaneous equations

$$\left.\begin{array}{r} y = x^3 \\ y = 48x - 128 \end{array}\right\} \implies x^3 = 48x - 128$$

This equation gives the x values of the points of intersection of curve and tangent: you know that $x = 4$ is one of the two points so $(x - 4)$ must be a factor of $x^3 - 48x + 128$.

$$\implies x^3 - 48x + 128 = 0$$

$$\implies (x - 4)(x^2 + 4x - 32) = 0$$

$$\implies (x - 4)(x + 8)(x - 4) = 0$$

$$\implies x = 4 \text{ or } -8$$

$$\implies Q(-8, -512).$$

You know $(x - 4)$ is a factor.

The second factor must have x^2 in it to obtain the x^3 term when the two brackets are multiplied out. The second factor must have -32 as the constant term to obtain the $+128$ when the two brackets are multiplied out.

You now have $(x - 4)(x^2 + bx - 32)$ as the factorisation: when this is multiplied out there will be two x^2 terms, $-4x^2$ and bx^2. These must add together to give $0x^2$ so the value of b must be 4.

c) Area between two curves

$$= \int_a^b (y_{top} - y_{bottom}) \, dx$$

$$= \int_{-8}^4 (x^3 - (48x - 128)) \, dx$$

$$= \int_{-8}^4 (x^3 - 48x + 128) \, dx$$

$$= \left[\frac{1}{4}x^4 - 24x^2 + 128x\right]_{-8}^4$$

$$= (64 - 384 + 512) - (1024 - 1536 - 1024)$$

$$= 192 - (-1536)$$

$$= 1728.$$

To summarise:

The area formed by the curves $y = f(x)$ and $y = g(x)$ and the lines $x = a$ and $x = b$ is

$$\int_a^b f(x) - g(x) \, dx$$

provided $f(x) \geqslant g(x)$ for all values of x between a and b.

Area Between a Curve and the y Axis

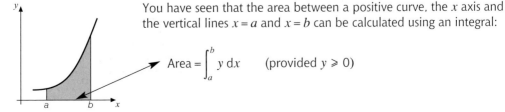

You have seen that the area between a positive curve, the x axis and the vertical lines $x = a$ and $x = b$ can be calculated using an integral:

$$\text{Area} = \int_a^b y \, dx \qquad \text{(provided } y \geqslant 0\text{)}$$

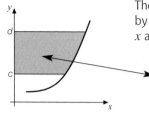

The area between a positive curve and the y axis can be calculated by calculating the integral obtained by simply switching the roles of x and y:

$$\text{Area} = \int_c^d x \, dy \qquad (\text{provided } x \geqslant 0)$$

EXAMPLE 19

The diagram shows the curve $y = x^2$ for $x > 0$.

a) Find the area of the closed region made by the curve, the lines $x = 2$, $x = 5$ and the x axis.

b) Find the area of the closed region made by the curve, the lines $y = 9$, $y = 16$ and the y axis.

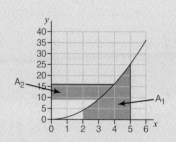

a) $A_1 = \int_2^5 y \, dx = \int_2^5 x^2 \, dx = \left[\dfrac{x^3}{3}\right]_2^5 = \dfrac{125}{3} - \dfrac{8}{3} = 39.$

b) To evaluate A_2 a formula for x in terms of y is needed.

$y = x^2$ and $x > 0 \Rightarrow x = \sqrt{y} = y^{1/2}$

so

$$A_2 = \int_9^{16} x \, dy = \int_9^{16} y^{1/2} \, dy = \left[\dfrac{y^{3/2}}{\frac{3}{2}}\right]_9^{16} = \dfrac{2}{3}\left[y^{3/2}\right]_9^{16} = \dfrac{2}{3}(64 - 27) = \dfrac{74}{3}.$$

EXERCISE 6

1 The diagram shows the graph of $y = x^2 - 4x - 5$.

Find the area of the shaded region bounded by the curve, the x axis and the lines $x = 2$ and $x = 6$.

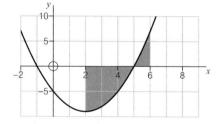

2 The diagram shows the curves

$y = x^2 - 9$ and $y = \dfrac{1}{10}(x^2 - 54)$ and

these curves intersect at $(-2, -5)$

and $(2, -5)$.

Calculate the area of the region between the two curves.

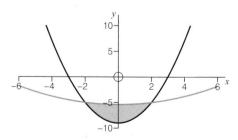

3 The diagram shows the graph of $y = \dfrac{30}{\sqrt{x}}$.

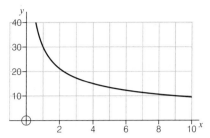

a) Calculate the area of the region bounded by the x axis, the lines $x = 1$ and $x = 9$ and the curve.

b) Calculate the area of the region bounded by the y axis, the lines $y = 15$ and $y = 30$ and the curve.

4 Sketch the graph of $y = 9x - x^3$.
Find the area of the region bounded by the curve, the x axis and the lines $x = 1$ and $x = 3$.

5 Sketch the graph of $y = x^4 - 16$.
Find the area of the closed region formed by the curve $y = x^4 - 16$ and the x axis.

6 Sketch on a single diagram the curves $y = x^2 + 6$ and $y = \dfrac{1}{2}x^3$ for values of x between 0 and 3.
Find the area of the region bounded by these curves and the lines $x = 0$ and $x = 2$.

7 Sketch the line $y = 4x$ and the curve $y = \dfrac{1}{2}x^2$.

Find the points of intersection of the line $y = 4x$ and the curve $y = \dfrac{1}{2}x^2$.

Find the area of the closed region formed by the line $y = 4x$ and the curve $y = \dfrac{1}{2}x^2$.

8 On a single diagram, sketch the graphs of $y = 9x^2$ and $y = x^4$ for positive and negative values of x.
Calculate the total area of the closed regions between the curves $y = 9x^2$ and $y = x^4$.

9 The diagram shows the curve $y = x^2$ and the normal to the curve at the point P(1, 1).

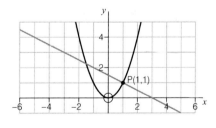

a) Find the equation of the normal.

b) Find the co-ordinates of the point Q where the normal meets the curve again.

c) Prove that the area of the closed region bounded by the curve and the normal is $\dfrac{125}{48}$.

10 a) Find the area of the closed region made by the curve $y = x^3$, the lines $x = 2$, $x = 4$ and the x axis.

b) Find the area of the closed region made by the curve $y = x^3$, the lines $y = 1$, $y = 8$ and the y axis.

11 a) Sketch the curve $y = \dfrac{36}{x^2}$ for $x > 0$.

b) Find the area of the closed region made by this curve, the lines $x = 2$, $x = 3$ and the x axis.

c) Find the area of the closed region made by this curve, the lines $y = 1$, $y = 4$ and the y axis.

The Trapezium Rule: An Approximate Means of Evaluating Integrals

You have seen that areas under curves correspond to definite integrals and that these can, in many cases, be evaluated exactly.

However, many definite integrals are either very hard or impossible to evaluate exactly.

You can, however, always revert to approximating the integral by a sum – that is approximating the area by a series of rectangular strips and this will give us a slow but sure way of getting an approximate answer for a definite integral or the area. The rectangular strip method is however inefficient – you will get much better approximations by approximating the area by a series of trapezia.

Consider for example the problem of finding the area under the graph of $y = x^2 + 2$ between $x = 1$ and $x = 3$.

You know the exact answer is $\int_1^3 x^2 + 2 \, dx = \left[\frac{1}{3}x^3 + 2x\right]_1^3 = 15 - \frac{7}{3} = \frac{38}{3}$.

One method of estimating the area (or integral) is to approximate the area by a series of rectangles:

area under graph ≈ area of rectangles
$$\approx 1 \times 3 + 1 \times 6 = 9.$$

This is clearly an underestimate of the required area.

A much better estimate can be obtained using two trapezia rather than two rectangles to approximate the area.

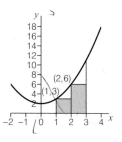

First recall the formula for the **area of a trapezium**:

area of trapezium = area of rectangle + area of triangle

$$= y_a h + \frac{1}{2} h(y_b - y_a)$$

$$= y_a h + \frac{1}{2} h y_b - \frac{1}{2} h y_a$$

$$= \frac{1}{2} y_a h + \frac{1}{2} y_b h$$

$$= \frac{1}{2} h(y_a + y_b).$$

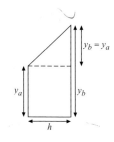

Area under curve ≈ total area of two trapezia

$$\approx \frac{1}{2} \times 1 \times (3 + 6) + \frac{1}{2} \times 1(6 + 11)$$

$$\approx 4.5 + 8.5 = 13 \text{ units}^2.$$

It can be seen that each trapezium slightly overestimates the actual area of the strip so this estimate will be an overestimate of the actual area under the curve.

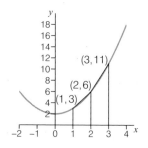

Recalling that the exact value of the area is $\dfrac{38}{3}$ and that approximation by rectangles gave an estimate of 9 for the area, it is clear that the estimate based on trapezia is much better than the approximation based on rectangles.

In general if you split a graph up into n trapezia, each of width h, and $y_0, y_1, ..., y_n$ are the ordinates then:

Area under curve $\approx$ total area under trapezia

$$\approx \frac{1}{2}h(y_0 + y_1) + \frac{1}{2}h(y_1 + y_2) + \cdots + \frac{1}{2}h(y_{n-1} + y_n)$$

$$\approx \frac{1}{2}h(y_0 + y_1 + y_1 + y_2 + \cdots + y_{n-1} + y_n)$$

$$\approx \frac{1}{2}h(y_0 + 2y_1 + 2y_2 + \cdots + 2y_{n-1} + y_n)$$

and this is the **trapezium rule** for estimating areas.

Area under curve $\approx \dfrac{1}{2}h(y_0 + 2y_1 + 2y_2 + \cdots + 2y_{n-1} + y_n)$

$\approx \dfrac{1}{2}h(y_0 + 2(y_1 + y_2 + \cdots + y_{n-1}) + y_n).$

EXAMPLE 20

Use the trapezium rule with eight trapezia to estimate the area of the region formed by the graph of $y = \dfrac{1}{x}$, the x axis and the lines $x = 1$ and $x = 5$.

The area under the graph between $x = 1$ and $x = 5$ is to be approximated by eight trapezia each of width 0.5.

$\dfrac{5-1}{8} = 0.5$

You need to calculate the nine y values required for the eight trapezia.

The table below shows these values (correct to three decimal places when necessary).

x	1	1.5	2	2.5	3	3.5	4	4.5	5
y	1	0.667	0.5	0.4	0.333	0.286	0.25	0.222	0.2

Area $\approx \dfrac{1}{2} \times 0.5 \times (1 + 2 \times 0.667 + 2 \times 0.5 + 2 \times 0.4 + 2 \times 0.333 + 2 \times 0.286 + 2$

$\times 0.25 + 2 \times 0.25 + 2 \times 0.222 + 0.2)$

≈ 1.63 units^2.

This will be an overestimate of the actual area since the top line of each trapezia is slightly above the corresponding part of the curve.

EXERCISE 7

1 Copy and complete the following table, giving values of $y = \dfrac{3}{\sqrt{1+x^3}}$ correct to three decimal places:

x	1	2	3	4	5
y			0.567		

Sketch the curve for values of x from 1 to 5.

By enclosing the curve between the lines $x = 1$ and $x = 5$ in a suitable trapezium, show that the area under the curve between the lines $x = 1$ and $x = 5$ is less than 4.8 units^2.

Use the trapezium rule with four strips to estimate this area.

2 Use the trapezium rule with six trapezia to estimate the value of $\displaystyle\int_0^3 \sqrt{4 + 2^x}\,dx$.

3 a) Explain why the curve

$$y = \sqrt{4 - x^2} \qquad x \geqslant 0$$

is a quarter circle and hence write down the exact value of $\displaystyle\int_0^2 \sqrt{4 - x^2}\,dx$.

b) Use the trapezium rule with five ordinates to estimate the value of

$$\int_0^2 \sqrt{4 - x^2}\,dx.$$

> Five ordinates means five y values: so four trapezia must be used.

c) Use a sketch graph to show how you can deduce that your answer to (b) is an underestimate of the value of π.

4 The velocity of a car starting from rest and moving in a straight line is recorded at 2-s intervals and the values are given in the table below:

t(s)	0	2	4	6	8	10	12
v(m/s)	0	8	15	20.5	25	28.5	30

Draw the graph of v against t, taking a scale of 1 cm to 1 s for t and 2 cm to 5 m/s for v. Use the trapezium rule to approximate the distance travelled by the car in the 12 s. Find also, from your graph, the acceleration of the car when $t = 6$ s.

Justification of the Calculation of Areas by Integration

Let $f(x)$ be a positive function.

Let $A(p)$ denote the area of the region bounded by the curve, the x axis and the lines $x = a$ and $x = p$.

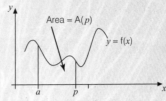

First of all show that $A'(p) = f(p)$.

Recall from C1 the formal definition

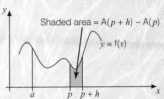

$$A'(p) = \lim_{h \to 0} \frac{A(p+h) - A(h)}{h}.$$

Now consider

$$
\begin{aligned}
A(p+h) - A(p) &= \text{area under curve between } x = a \text{ and } x = p + h \\
&\quad - \text{area under curve between } x = a \text{ and } x = p \\
&= \text{area under curve between } x = p \text{ and } x = p + h.
\end{aligned}
$$

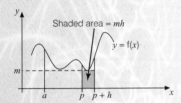

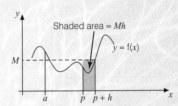

If m denotes the **minimum** value of $f(x)$ for values of x between p and $p + h$ and M denotes the **maximum** value of $f(x)$ for values of x between p and $p + h$ the diagrams show that

$$mh \leqslant A(p+h) - A(p) \leqslant Mh.$$

Dividing through by h gives

$$m \leqslant \frac{A(p+h) - A(p)}{h} \leqslant M.$$

If you now let $h \to 0$ then

$$\frac{A(p+h) - A(p)}{h} \to A'(p).$$

If the function f is continuous (i.e. its graph has no jumps in it) then the minimum value, m, of $f(x)$ for values of x between p and $p+h$ will get closer and closer to $f(p)$ as $h \to 0$. The maximum value, M of $f(x)$ for values of x between p and $p+h$ will also get closer and closer to $f(p)$ as $h \to 0$.

Letting $h \to 0$, the inequality

$$m \leqslant \frac{A(p+h) - A(p)}{h} \leqslant M$$

becomes

$$f(p) \leqslant A'(p) \leqslant f(p)$$
$$\Rightarrow \quad A'(p) = f(p).$$

The derivative of the **area** function is the original function, f(x).

The area function can therefore be found by integrating the original function, f(x).

You have proved that integrating the curve equation gives areas.

Justification of the Result for the Area Between Two Curves

Consider the area between the curves $y = f(x)$ and
$y = g(x)$ and the lines $x = a$ and $x = b$, where
$f(x) \geqslant g(x)$ for all values of x between $x = a$ and $x = b$.

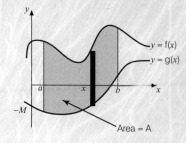

Consider the small rectangular strip of width δx shown
in black on the diagram.
The height of the strip is $f(x_i) - g(x_i)$ so the area of the
strip is $(f(x_i) - g(x_i))\,\delta x$ and by adding up the areas of
lots of these strips you obtain

$$A \approx \sum_{x=a}^{b} (f(x_i) - g(x_i))\,\delta x.$$

To improve the estimate you need to take lots of very thin strips so we can write

$$A = \lim_{\delta x \to 0} \sum_{x=a}^{b} (f(x_i) - g(x_i))\,\delta x$$

and, from your earlier work, you know that the limit of sums like this can be written as
integrals so you have

$$A = \int_a^b (f(x) - g(x))\,dx.$$

Having studied this chapter you should know

- that integration is the reverse of differentiation and be able to use the result

$$\int x^n \, dx = \frac{x^{n+1}}{n+1} + c \qquad (n \neq -1)$$

- that if f(x) is positive then the area of the region formed by the curve $y = f(x)$,
the x axis and the lines $x = a$ and $x = b$ is given by the definite integral

$$\int_a^b f(x)\,dx = \int_a^b y\,dx$$

- that if f(x) is negative between $x = a$ and $x = b$ then $\int_a^b f(x)\,dx$ will give a negative
answer and the area of the region formed by the curve $y = f(x)$, the x axis and
the lines $x = a$ and $x = b$ is given by $-\int_a^b f(x)\,dx$

- that the area formed by the curves $y = f(x)$ and $y = g(x)$ and the lines $x = a$ and $x = b$ is

$$\int_a^b (f(x) - g(x))\, dx$$

provided $f(x) \geqslant g(x)$ for all values of x between a and b

- that if $g(y)$ is positive then the area of the region formed by the curve $x = g(y)$, the y axis and the lines $y = c$ and $y = d$ is given by the definite integral

$$\int_c^d g(y)\, dy = \int_c^d x\, dy$$

- to use the trapezium rule to estimate definite integrals and areas under graphs:

area under curve $\approx \dfrac{1}{2} h(y_0 + 2y_1 + 2y_2 + \cdots + 2y_{n-1} + y_n)$

$$\approx \frac{1}{2} h(y_0 + 2(y_1 + y_2 + \cdots + y_{n-1}) + y_n)$$

REVISION EXERCISE

1 Evaluate

a) $\displaystyle\int_0^2 x(x + 2)\, dx$ **b)** $\displaystyle\int_1^4 \frac{4}{x^2}\, dx$

2 The gradient at the point (x, y) on a curve is given by

$$\frac{dy}{dx} = x^2(2x + 3).$$

The curve passes through the point $(0, 2)$.
a) Find the equation of the curve.
b) Find the area of the closed region formed by the curve, the x axis and the lines $x = 1$ and $x = 3$.

3 Sketch the curve $y = x^5$.
Find the area of the closed region bounded by the curve, the y axis and the lines $y = 1$ and $y = 32$.

4 **a)** On a single diagram, sketch the graphs of $y = \dfrac{1}{2} x^2$ and $y = 2x + 6$.

     b) Find the points of intersection of $y = \dfrac{1}{2} x^2$ and $y = 2x + 6$.

     c) Find the area of the closed region bounded by the curves $y = \dfrac{1}{2} x^2$ and $y = 2x + 6$.

5 **a)** Sketch the graph of $y = x^2(x - 1)$.
     b) Find the equation of the tangent to this curve at the point $P(1, 0)$ and show the tangent on your sketch.
     c) Verify that the tangent and the curve intersect at the point $Q(-1, -2)$.
     d) Find the area of the closed region bounded by the curve and the line PQ.

6 Evaluate

a) $\displaystyle\int_4^9 \sqrt{t}\, dt$ b) $\displaystyle\int_1^2 \frac{t+2}{t^3}\, dt$

7 a) Find the area of the region bounded by the curve $y = \sqrt[4]{x}$, the x axis and the lines $x = 1$ and $x = 16$.

b) Find the area of the region bounded by the curve $y = \sqrt[4]{x}$, the y axis and the lines $y = 1$ and $y = 3$.

8 Use the trapezium rule with three intervals to estimate the value of

$$\int_0^6 \sqrt{4 + 3x}\, dx.$$

9 Evaluate

a) $\displaystyle\int_0^{32} \frac{4}{\sqrt[5]{x}}\, dx$ b) $\displaystyle\int_4^{\infty} \frac{10}{t^{3.5}}\, dt$

10 Find $\displaystyle\int \left(2x^3 - 3x + \frac{6}{x^2}\right) dx.$

11 a) The gradient of a curve, C_1, at the point (x, y) is given by

$$\frac{dy}{dx} = 5 - 2x$$

and the curve passes through the point $(2, 3)$. Find the equation of the curve C_1.

b) The curve C_2 has equation $y = x^2 - 1$. Find the points of intersection of C_1 and C_2.

c) Sketch the curves C_1 and C_2 on a single diagram and show that the area of the closed region bounded by the two curves is given by

$$\int_{\frac{1}{2}}^2 5x - 2 - 2x^2 \, dx.$$

d) Hence show that the area of the closed region bounded by the two curves is $\dfrac{9}{8}$.

12 The diagram shows the graph of $y = \sqrt{9 - 3 \times 2^{-x}}$.

The region R, shaded in the diagram, is bounded by the curve and by the lines $x = -1$, $x = 1$ and $y = 0$.

a) Use the trapezium rule with four intervals to find an estimate for the area of R.

b) State, with a reason, whether the trapezium rule gives an underestimate or an overestimate of the area of R.

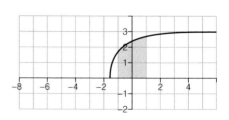

6 Exponentials and Logarithms

The purpose of this chapter is to enable you to

- recognise the exponential function and appreciate its use in modelling growth and decay
- recognise and use the logarithm function and its fundamental properties

The Exponential Function

An exponential function is any function whose rule takes the form $f(x) = a^x$ where a is a positive constant.
The diagram shows the graphs of $y = 1.6^x$ and $y = 2.8^x$.

Notice that in both cases

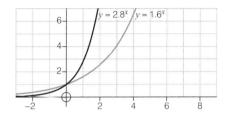

- the graph takes only positive values
- the graph passes through $(0, 1)$
- as $x \rightarrow \infty$, y gets large very quickly
- as $x \rightarrow -\infty$, $y \rightarrow 0$.

The second diagram shows the graphs of $y = 0.75^x$ and $y = 0.4^x$.

Notice that in both cases

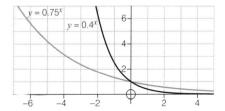

- the graph takes only positive values
- the graph passes through $(0, 1)$
- as $x \rightarrow \infty$, $y \rightarrow 0$
- as $x \rightarrow -\infty$, y gets large very quickly.

In general,

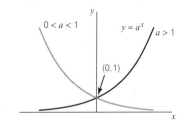

- the graph of $y = a^x$ will pass through $(0, 1)$ since a^0 is always 1
- will be increasing if $a > 1$
- will be decreasing if $0 < a < 1$.

If the graph of $y = a^x$ is stretched in the y direction by scale factor k then the image has equation $y = k \times a^x$ and, whatever the value of a, this will pass through the point $(0, k)$.

The diagram shows typical graphs of

$$y = ka^x$$

for different values of a.

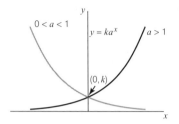

If $a > 1$ then the graph is increasing and the curve is an example of **an exponential growth curve**.

If $0 < a < 1$ then the graph is decreasing and the curve is **an exponential decay curve**.

EXAMPLE 1

The table below shows a predicted value, £V, in t months time of a car bought initially for £14 500.

t	0	6	12	18	24	36	48	60	72	84
V	14 500	13 100	11 700	10 600	9500	7700	6200	5100	4100	3300

a) Plot a scatter graph for this data and obtain a rule linking V and t.
b) Hence predict the value of this car 20 months after it is purchased.

a) On the scatter diagram of V against t, the points look as if they lie on an **exponential decay curve** so it is reasonable to expect the data to follow a relationship of the form

$$V = k \times a^t \quad \text{where} \quad 0 < a < 1.$$

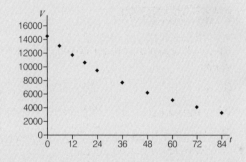

You have (0, 14 500) as a data point and you know that the curve $V = ka^t$ passes through the point (0, k) so it seems reasonable to proceed by putting $k = 14\,500$.

You can therefore try to fit a rule of the form $V = 14\,500 \times a^t$ to the data.

To find the value of a, use one of the data points: when $t = 24$; $V = 9500$.

Substituting these values into the equation, $V = 14\,500 \times a^t$, gives

$$9500 = 14\,500 \times a^{24}$$

$$\Rightarrow \quad a^{24} = \frac{9500}{14\,500}$$

$$\Rightarrow \quad a = \left(\frac{9500}{14\,500}\right)^{1/24} = 0.9825.$$

So the rule appears to be $V = 14\,500 \times 0.9825^t$.

EXAMPLE 1 (continued)

The reliability of this rule can be checked by calculating the values of $14\,500 \times 0.9825^t$, correct to the nearest pound, and adding them to the table of results and by adding the curve $V = 14\,500 \times 0.9825^t$ to the scatter diagram:

t	0	6	12	18	24	36	48	60	72	84
V	14 500	13 100	11 700	10 600	9500	7700	6200	5100	4100	3300
$14\,500 \times 0.9825^t$	14 500	13 043	11 732	10 552	9492	7680	6213	5027	4067	3291

It can be seen from the table, and the graph, that the rule $V = 14\,500 \times 0.9825^t$ does accurately describe the data given in the table.

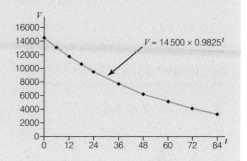

b) Putting $t = 20$ into the rule, you obtain a predicted value of $14\,500 \times 0.9825^{20} = £10\,200$, to the nearest £100.

EXERCISE 1

Sketch the graphs of the following functions, taking care to show where the graphs cross the y axis:

1 $y = 2.4^x$ **2** $y = 3.1^x$ **3** $y = 0.92^x$ **4** $y = 5 \times 2^x$

5 $y = 20 \times 1.3^x$ **6** $y = 30 \times 0.4^x$ **7** $y = 80 \times 0.7^x$ **8** $y = 50 \times 3.2^x$

9 The table below shows the population, N people, of a new town t months after the new town was formally opened:

t	0	4	8	12	16	20	24
N	20 000	22 050	24 310	26 802	29 549	32 578	35 917

t	28	32	36	40	44	48
N	39 599	43 657	48 132	53 066	58 505	64 502

a) Draw a scatter diagram to illustrate this data.
b) Find a suitable rule for the data and show the curve for this rule on your diagram.
c) Use your answer to (b) to obtain an estimate for the population of the new town 18 months after it was formally opened.
d) Explain why it would not be sensible to assume that this rule linking the population and time will continue indefinitely.

10 James has a pocket computer. The batteries of this computer need recharging regularly. The percentage of full charge, $P\%$, remaining after the computer has been left turned off for t hours is given in the table below:

t	0	12	24	36	48	72	96	120	144	168
P	100	83	70	58	48	34	23	16	11	8

a) Draw a scatter diagram to illustrate this data.
b) Find a suitable rule for the data and show the curve for this rule on your diagram.
c) Use your answer to (b) to obtain an estimate for percentage of full charge remaining after 50 hours.

The Logarithm Function

The Rules of Indices

To understand logarithms you must thoroughly understand the use and rules of indices:

- $a^m \times a^n = a^{m+n}$
- $(a^m)^p = a^{mp}$
- $a^{-n} = \dfrac{1}{a^n}$
- $a^{q/p} = (a^{1/p})^q = (\sqrt[p]{a})^q$
- $\left(\dfrac{p}{q}\right)^n = \dfrac{p^n}{q^n}$

- $a^m \div a^n = a^{m-n}$
- $a^0 = 1$
- $a^{1/p} = \sqrt[p]{a}$
- $(pq)^n = p^n q^n$

EXERCISE 2

> Make sure you can do these questions before progressing to the next section. If necessary look back at the C1 work on indices to revise the topic thoroughly.

1 Simplify the following:

a) $3^5 \times 3^2 \times 3^7$　　**b)** $p^2 \times p^5 \times p^8$　　**c)** $\dfrac{5^7}{5^3}$　　**d)** $\dfrac{2^3 \times 2^4 \times 2^7}{2^2 \times 2^5}$

e) $(7^2)^3$　　**f)** $(3x^2 y^4)^5$　　**g)** $\left(\dfrac{5a^2}{2b^3}\right)^3$　　**h)** $(2a^2 b^3 c)^4$

i) $(16x^6)^{1/2}$　　**j)** $(64x^6)^{1/3}$　　**k)** $(4x^3)^{-2}$　　**l)** $(2x^5)^{-3}$

m) $(3x^2)^{-4}$　　**n)** $(8x^6)^{2/3}$　　**o)** $(81x^8)^{3/4}$　　**p)** $(4x^8)^{-1/2}$

q) $(9x^8)^{-1/2}$　　**q)** $(125x^6)^{-2/3}$　　**s)** $(32x^{10})^{2/5}$　　**t)** $(81x^8 y^{12})^{-3/4}$

2 Express the following numbers as powers of 3:

a) 9^2　　**b)** 27^4　　**c)** 81^{-3}　　**d)** $\dfrac{1}{243}$　　**e)** $\dfrac{1}{\sqrt{243}}$

3 Find the value of p if

a) $3^p = 9$　　**b)** $3^{2p-1} = 243$　　**c)** $2^{p+1} = 4^{p-2}$　　**d)** $5^{2p+3} = 125^{p+5}$

Logarithms: The Basic Idea

A statement involving logarithms is simply a different way of writing a statement involving indices.

> For positive numbers x and a
> if $\quad x = a^p$
> then p is called the logarithm to base a of x and this is written
> $\qquad \log_a x = p.$

This definition can be used to write down the values of many logarithms.

EXAMPLE 2

Find the values of

a) $\log_3 81$ b) $\log_4 \left(\dfrac{1}{64} \right)$ c) $\log_8 2$

SOLUTION

a) The statement $81 = 3^4$ can be rewritten as $\log_3 81 = 4$.

b) Since $\dfrac{1}{64} = \dfrac{1}{4^3} = 4^{-3}$ you can say that $\log_4 \left(\dfrac{1}{64} \right) = -3$.

c) Since $2 = \sqrt[3]{8} = 8^{1/3}$ you can also say that $\log_8 2 = \dfrac{1}{3}$.

The definition of logarithms can be used in reverse to solve equations involving logarithms.

EXAMPLE 3

Solve the equations

a) $\log_6 t = 3$ b) $\log_{16} y = -\dfrac{1}{2}$ c) $\log_4(5z - 1) = 3$

SOLUTION

a) $\log_6 t = 3 \qquad \Rightarrow \quad t = 6^3 = 216$

b) $\log_{16} y = -\dfrac{1}{2} \qquad \Rightarrow \quad y = 16^{-1/2} = \dfrac{1}{16^{1/2}} = \dfrac{1}{\sqrt{16}} = \dfrac{1}{4}$

> The key step in solving these equations is remembering that the statement $\log_a x = p$ is equivalent to the statement $x = a^p$.

c) $\log_4(5z - 1) = 3 \quad \Rightarrow \quad 5z - 1 = 4^3 = 64$
$\qquad\qquad\qquad\qquad \Rightarrow \quad 5z = 65$
$\qquad\qquad\qquad\qquad \Rightarrow \quad z = 13.$

Using Trial and Improvement Methods to Find the Values of Logarithms

The values $\log_2 4 = 2$ and $\log_2 8 = 3$ follow immediately from the statements $4 = 2^2$ and $8 = 2^3$. However, the value of $\log_2 6$ is not obvious since there is no obvious value p satisfying the equation $2^p = 6$ but a solution can be found by a **trial and improvement** method.

Suppose $\log_2 6 = p \Rightarrow 6 = 2^p$.

You know that $2^2 = 4$ and $2^3 = 8$ so p lies between 2 and 3.

Using trial and improvement, you obtain:

p	2^p
2	4
3	8
2.5	5.65 ...
2.6	6.06 ...
2.58	5.97 ...
2.585	6.0001 ...

From this table you can see that p lies between 2.58 and 2.585 so we can be certain that $p = 2.58$, correct to two decimal places.

The statement "the solution of $2^p = 6$ is $p = 2.58$, correct to two decimal places" is equivalent to the statement "the value of $\log_2 6$ is 2.58, correct to two decimal places".

Later in this chapter a much quicker means of evaluating logarithms of numbers will be demonstrated.

The Graph of the Logarithm Function

You know that

$\dfrac{1}{4} = 2^{-2}$ and that the equivalent statement involving logarithms is $\log_2 \dfrac{1}{4} = -2$,

$\dfrac{1}{2} = 2^{-1}$ and that the equivalent statement involving logarithms is $\log_2 \dfrac{1}{2} = -1$,

$1 = 2^0$ and that the equivalent statement involving logarithms is $\log_2 1 = 0$,
$2 = 2^1$ and that the equivalent statement involving logarithms is $\log_2 1 = 1$,
$4 = 2^2$ and that the equivalent statement involving logarithms is $\log_2 4 = 2$,
$8 = 2^3$ and that the equivalent statement involving logarithms is $\log_2 8 = 3$.

These facts enable you to draw the graph of $y = \log_2 x$.

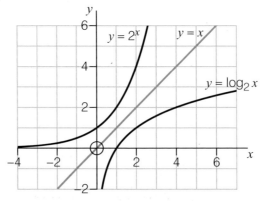

Notice that the graph of $y = \log_2 x$ is the image of $y = 2^x$ after a reflection in the line $y = x$.

EXERCISE 3

1 Write down the values of

a) $\log_3 9$ **b)** $\log_3 27$ **c)** $\log_3 \dfrac{1}{9}$ **d)** $\log_3 \sqrt{3}$ **e)** $\log_3 \dfrac{1}{\sqrt{3}}$

f) $\log_{10} 1000$ **g)** $\log_{10} 100$ **h)** $\log_{10} 10$ **i)** $\log_{10} 1$ **j)** $\log_{10} \dfrac{1}{100}$

k) $\log_5 125$ **l)** $\log_5 \dfrac{1}{25}$ **m)** $\log_5 \sqrt{5}$ **n)** $\log_5 5\sqrt{5}$ **o)** $\log_5 \dfrac{1}{\sqrt{5}}$

p) $\log_7 1$ **q)** $\log_7 49$ **r)** $\log_7 \dfrac{1}{\sqrt{7}}$ **s)** $\log_7 \dfrac{1}{49}$

2 Find the value of x if

a) $\log_2 x = 3$ **b)** $\log_4 x = 2.5$ **c)** $\log_5 x = -1$

d) $\log_5 x = 2$ **e)** $\log_9 x = \dfrac{1}{2}$ **f)** $\log_{16} x = -\dfrac{1}{2}$

g) $\log_2(x + 4) = 4$ **h)** $\log_7(3x + 1) = 2$ **i)** $\log_5(2x + 1) = \dfrac{1}{2}$

3 Use a trial and improvement method to determine the value, correct to three decimal places, of $\log_4 12$.

4 a) Draw the graph of $y = 3^x$ for $-2 \leqslant x \leqslant 3$.

   b) Complete the table below and hence draw on the same diagram the graph of $y = \log_3 x$:

x	$\dfrac{1}{9}$	$\dfrac{1}{3}$	1	3	9	27
$y = \log_3 x$						

   c) What transformation will map the graph of $y = 3^x$ onto the graph of $y = \log_3 x$?

5 Write down the values of

a) $\log_2 4$, $\log_2 8$ and $\log_2 32$;

b) $\log_{16} 4$, $\log_{16} 16$ and $\log_{16} 64$;

c) $\log_{10} \sqrt{10}$, $\log_{10} 100$ and $\log_{10}(100\sqrt{10})$;

d) $\log_5 \dfrac{1}{5}$, $\log_5 125$ and $\log_5 25$.

> Questions 5 and 6 introduce some of the properties of logarithms that will be met in the next section.

What connection do these results suggest between $\log_a x$, $\log_a y$ and $\log_a xy$?

6 Copy and complete the following table (use your calculator to obtain the values of $\log x$ which is the normal shorthand for $\log_{10} x$):

x	0.25	0.5	1	2	4	16	64
$\log_4 x$							
$\log_{10} x$							

> $\log_{10} x$ is usually written as $\log x$. Values of $\log x$ can be obtained easily from your calculator.

Can you find a rule connecting $\log_4 x$ and $\log_{10} x$?

Properties of Logarithms

In questions 5 and 6 of Exercise 3 some of the basic properties of logarithms were introduced. In this section six rules of logarithms will be introduced and in each case the proof of the rule is a consequence of the definition of logarithms and the properties of indices.

You know that $\log_3 9 = 2$ and $\log_3 27 = 3$ since $9 = 3^2$ and $27 = 3^3$.

You also know that $\log_3(9 \times 27) = \log_3(243) = 5$ since $243 = 3^5$.

Putting these results together you have

$$\log_3(9 \times 27) = 5 = 2 + 3 = \log_3 9 + \log_3 27.$$

This can be generalised to:

> **Rule 1** $\log_a(xy) = \log_a x + \log_a y$

Proof:
Suppose $\log_a x = p$ and $\log_a y = q$.

Then

$x = a^p$ and $y = a^q$

> Remember the defining property of a logarithm: writing $\log_a x = p$ means that $x = a^p$.

so

$$xy = a^p \times a^q = a^{p+q} \quad \text{(rules of indices)}.$$

Now

$$xy = a^{p+q} \implies \log_a(xy) = p + q \implies \log_a(xy) = \log_a x + \log_a y.$$

You know that $\log_3(9 \div 27) = \log_3\left(\frac{1}{3}\right) = -1$ since $\frac{1}{3} = 3^{-1}$.

You therefore have

$$\log_3(9 \div 27) = -1 = 2 - 3 = \log_3 9 - \log_3 27.$$

This can be generalised to

> **Rule 2** $\log_a\left(\dfrac{x}{y}\right) = \log_a x - \log_a y$

Proof:

Suppose $\log_a x = p$ and $\log_a y = q$.

Then $x = a^p$ and $y = a^q$

So

$$\frac{x}{y} = a^p \div a^q = a^{p-q} \quad \text{(rules of indices).}$$

Now

$$\frac{x}{y} = a^{p-q} \quad \Rightarrow \quad \log_a\left(\frac{x}{y}\right) = p - q \quad \Rightarrow \quad \log_a\left(\frac{x}{y}\right) = \log_a x - \log_a y.$$

You know that $\log_2 4 = 2$ since $4 = 2^2$.

You also know that $\log_2(4^3) = \log_2(64) = 6$ since $64 = 2^6$.

You therefore have

$$\log_2(4^3) = 6 = 3 \times 2 = 3 \times \log_2 4$$

which is a special case of

> **Rule 3** $\log_a(x^n) = n \log_a x$

Proof:

Suppose $\log_a x = p$.

Then

$$x = a^p$$

so

$$x^n = (a^p)^n = a^{np} \quad \text{(rules of indices).}$$

Now

$$x^n = a^{np} \quad \Rightarrow \quad \log_a(x^n) = np \quad \Rightarrow \quad \log_a(x^n) = n \log_a x.$$

You know from your work on indices that, for any positive number a, $1 = a^0$. The logarithm form of this statement is

> **Rule 4** $\log_a(1) = 0$

You also know that $a = a^1$ and the logarithm form of this statement is

> **Rule 5** $\log_a(a) = 1$

EXTENSION

The following rule shows how the logarithm, to any base, of any positive number can be easily obtained using your calculator.

The specification for the C2 module does **not** require knowledge of this rule.

Rule 6 $\log_a x = \dfrac{\log_{10} x}{\log_{10} a}$

It is usual to write $\log_{10} x$ as $\log x$ so this rule is usually written as $\log_a x = \dfrac{\log x}{\log a}$.

Proof:

Suppose $\log_a x = p$.

Then

$$x = a^p$$

so

Using rule 3

$$\log_{10} x = \log_{10}(a^p)$$

$$\Rightarrow \quad \log_{10} x = p \log_{10} a$$

$$\Rightarrow \quad p = \frac{\log_{10} x}{\log_{10} a}$$

$$\Rightarrow \quad \log_a x = \frac{\log_{10} x}{\log_{10} a}.$$

Earlier in the chapter a trial and improvement method was used to discover that

$$\log_2 6 = 2.58 \quad \text{(correct to two decimal places)}.$$

Rule 6 can now be used to write down the value of $\log_2 6$:

$$\log_2 6 = \frac{\log 6}{\log 2} = 2.58496\ldots$$

EXAMPLE 4

Simplify the expressions

a) $\log_4 25 - 2 \log_4 10$ **b)** $\log_7 16 \div \log_7 2$

Using rule 3

a) $\log_4 25 - 2 \log_4 10 = \log_4 25 - \log_4(10^2)$

$$= \log_4 25 - \log_4 100$$

Using rule 2

$$= \log_4\left(\frac{25}{100}\right)$$

$\dfrac{1}{4} = 4^{-1}$ so $\log_4\left(\dfrac{1}{4}\right) = -1$.

$$= \log_4\left(\frac{1}{4}\right)$$

$$= -1.$$

EXAMPLE 4 (continued)

b) $\log_7 16 \div \log_7 2 = \dfrac{\log_7 16}{\log_7 2}$

$$= \frac{\log_7(2^4)}{\log_7 2}$$

> Using rule 3

$$= \frac{4 \log_7 2}{\log_7 2}$$

$$= 4.$$

EXAMPLE 5

Find the value of 3245^{2005}, giving your answer in standard form correct to three significant figures.

Let $y = 3245^{2005}$

> See what happens if you try to evaluate it on your calculator using the power button!

then

$$\log y = \log(3245^{2005})$$

> Using rule 3

$$= 2005 \log 3245$$

$$\approx 7039.985476 \ldots$$

so

> Remember $\log_{10} y = p$ means that $y = 10^p$.

$$y = 10^{7039.985476 \ldots}$$

$$= 10^{0.985476 \ldots} \times 10^{7039}$$

$$= 9.671 \ldots \times 10^{7039}$$

so

$$3245^{2005} = 9.67 \times 10^{7039}, \text{ correct to three significant figures.}$$

Using Logarithms to Solve Equations Involving Indices

Logarithms provide an important tool that can be used to solve a wide range of equations involving indices.

EXAMPLE 6

Solve the equation $4^p = 57$ giving your answer to three decimal places.

$$4^p = 57$$

> Using rule 3

$$\Rightarrow \quad \log(4^p) = \log 57$$

$$\Rightarrow \quad p \log 4 = \log 57$$

$$\Rightarrow \quad p = \frac{\log 57}{\log 4} \approx 2.916 \quad (3 \text{ d.p.}).$$

EXAMPLE 7

Prove that the solution of the equation $5^{t+1} = 3^{2t-1}$ is $t = \dfrac{\log 15}{\log 1.8}$.

SOLUTION

$$5^{t+1} = 3^{2t-1}$$

$\Longrightarrow \quad \log(5^{t+1}) = \log(3^{2t-1})$ — Using rule 3

$\Longrightarrow \quad (t+1)\log 5 = (2t-1)\log 3$

$\Longrightarrow \quad t\log 5 + \log 5 = 2t\log 3 - \log 3$

$\Longrightarrow \quad \log 5 + \log 3 = 2t\log 3 - t\log 5$ — Using rule 1 on the left hand side and rule 3 on the right hand side.

$\Longrightarrow \quad \log 5 + \log 3 = t(2\log 3 - \log 5)$

$\Longrightarrow \quad \log(5 \times 3) = t(\log(3^2) - \log 5)$

$\Longrightarrow \quad \log 15 = t(\log 9 - \log 5)$ — Using rule 2

$\Longrightarrow \quad \log 15 = t\log\left(\dfrac{9}{5}\right)$

$\Longrightarrow \quad t = \dfrac{\log 15}{\log 1.8}$

EXAMPLE 8

Solve, correct to three decimal places, the equation $5^{2x} - 16 \times 5^x + 63 = 0$.

SOLUTION

You met equations similar to this in chapter 8 of C1. The key observation is that

$$5^{2x} = (5^x)^2$$

and this means that the equation can be regarded as a quadratic equation with 5^x as the variable.

$$5^{2x} - 16 \times 5^x + 63 = 0$$

$\Longrightarrow \quad (5^x)^2 - 16 \times 5^x + 63 = 0.$

This is a quadratic in 5^x which can be factorised:

$\Longrightarrow \quad (5^x - 7)(5^x - 9) = 0$

$\Longrightarrow \quad 5^x = 7 \quad \text{or} \quad 5^x = 9$

$\Longrightarrow \quad \log(5^x) = \log 7 \quad \text{or} \quad \log(5^x) = \log 9$ — Using rule 3

$\Longrightarrow \quad x\log 5 = \log 7 \quad \text{or} \quad x\log 5 = \log 9$

$\Longrightarrow \quad x = \dfrac{\log 7}{\log 5} \quad \text{or} \quad x = \dfrac{\log 9}{\log 5}$

$\Longrightarrow \quad x = 1.209 \ (3 \text{ d.p.}) \quad \text{or} \quad x = 1.365 \quad (3 \text{ d.p.}).$

EXAMPLE 9

Solve the equation $3^{2x+1} - 12 \times 3^x - 57 = 0$.

$$3^{2x+1} - 12 \times 3^x - 57 = 0$$

Using the rules of indices:
$$3^{2x+1} = 3^{2x} \times 3^1 = (3^x)^2 \times 3 = 3 \times (3^x)^2.$$

$$\Rightarrow \quad 3 \times (3^x)^2 - 12 \times 3^x - 57 = 0.$$

Viewing this as a quadratic in 3^x gives

This time, the quadratic will not factorise so the quadratic equation formula must be used.

$$3^x = \frac{12 \pm \sqrt{(-12)^2 - 4 \times 3 \times (-57)}}{2 \times 3}$$

$$= 6.7958 \ldots \text{ or } -2.7958 \ldots$$

3^x is always a positive number so the $-2.7958 \ldots$ solution can be rejected:

$$3^x = 6.7958 \ldots$$

$$\Rightarrow \quad \log(3^x) = \log(6.7958 \ldots)$$

$$\Rightarrow \quad x \log 3 = \log(6.7958 \ldots)$$

$$\Rightarrow \quad x = \frac{\log(6.7958 \ldots)}{\log 3} = 1.744. \quad \text{(to 3 d.p.)}$$

Equations involving indices arise naturally from practical situations of exponential growth and decay such as the growth of the values of investments or the depreciation of the value of a car.

EXAMPLE 10

Assuming that the price of a house increases at a rate of 12% per annum, how long will it take a house currently valued at £280 000 to reach a value of £1 000 000?

Each year the value of the house increases by 12%. In other words the value of the house in 12 months' time will be 112% of its present price or 1.12 times its present price.

If the house is currently valued at £280 000

- In a year's time it will be worth $280\,000 \times 1.12$;
- In 2 years' time it will be worth $280\,000 \times 1.12^2$;
- In 3 years' time it will be worth $280\,000 \times 1.12^3$;
- In n years' time it will be worth $280\,000 \times 1.12^n$.

You want

$$280\,000 \times 1.12^n = 1\,000\,000$$

$$\Rightarrow \quad 1.12^n = 3.5714 \ldots$$

Using rule 3

$$\Rightarrow \quad \log(1.12^n) = \log(3.5714 \ldots)$$

$$\Rightarrow \quad n \log 1.12 = \log(3.5714 \ldots)$$

$$\Rightarrow \quad n = 11.23 \quad \text{(2 d.p.)}.$$

So the house will be worth £1 000 000 after approximately $11\frac{1}{4}$ years.

Using Logarithms to Solve Inequalities Involving Indices

Inequalities involving indices can also be solved with the aid of logarithms. The solution will frequently involve dividing each side of the inequality by a logarithm. You will see in example 11 that it is important to think about whether this logarithm will be a positive or negative number since if the value of the logarithm is negative then dividing each side of the inequality by the logarithm will reverse the direction of the inequality.

EXAMPLE 11

Solve the inequalities **a)** $1.2^x < 5$
b) $0.4^y < 0.0001$

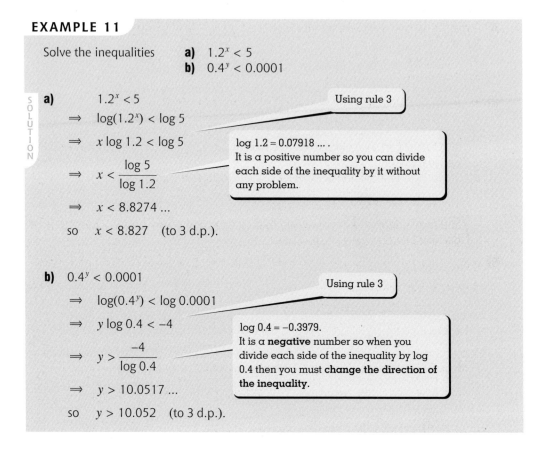

a) $1.2^x < 5$

$\Rightarrow \quad \log(1.2^x) < \log 5$ — Using rule 3

$\Rightarrow \quad x \log 1.2 < \log 5$

$\log 1.2 = 0.07918 \dots$.
It is a positive number so you can divide each side of the inequality by it without any problem.

$\Rightarrow \quad x < \dfrac{\log 5}{\log 1.2}$

$\Rightarrow \quad x < 8.8274 \dots$

so $\quad x < 8.827 \quad$ (to 3 d.p.).

b) $0.4^y < 0.0001$

$\Rightarrow \quad \log(0.4^y) < \log 0.0001$ — Using rule 3

$\Rightarrow \quad y \log 0.4 < -4$

$\log 0.4 = -0.3979$.
It is a **negative** number so when you divide each side of the inequality by log 0.4 then you must **change the direction of the inequality.**

$\Rightarrow \quad y > \dfrac{-4}{\log 0.4}$

$\Rightarrow \quad y > 10.0517 \dots$

so $\quad y > 10.052 \quad$ (to 3 d.p.).

EXERCISE 4

1 Without using tables or a calculator, calculate or simplify the following:

a) $\log_{10} 30 - \log_{10} 3$

b) $\log_{10} 8 + \log_{10} 25 - \log_{10} 2$

c) $\log_2 16 - \log_2 4$

d) $\log_3 2.7 + \log_3 10$

e) $\log_{10} 8 \div \log_{10} 2$

f) $\log_{10} 32 \div \log_{10} 2$

g) $\log_a a\sqrt{a}$

h) $\log_{10} 7 + 2\log_{10} 5 - \log_{10}\left(\dfrac{7}{4}\right)$

i) $\log_6 2 - \log_6 12$

j) $\log_6 42 - \log_6 7$

k) $\log_3 39 + \log_3 3 - \log_3 13$

l) $\log_a a^5 - \log_a \sqrt{a}$

m) $\log_4 1.6 + \log_4 10$

n) $\log_8 72 - \log_8 18 + \log_8 2$

2 Solve the following equations, giving your answers correct to three decimal places:

a) $3^x = 29$

b) $5^x = 0.3$

c) $12^x = 17$

d) $4^{2x-1} = 100$

e) $0.8^{3x-1} = 0.01$

f) $3^t = 2^{t+1}$

3 Find the values of

a) $\log_4 7$

b) $\log_7 877$

c) $\log_6 500$

d) $\log_8 0.0012$

4 Solve the following equations, giving your answers correct to three decimal places, where appropriate:

a) $2^{2x} - 5 \times 2^x + 4 = 0$

b) $3^{2x+1} - 7 \times 3^x + 2 = 0$

c) $7^{2x+1} - 3 \times 7^{x+1} - 240 = 0$

5 a) Show that the equation $16^x - 5 \times 4^{x+1} + 84 = 0$ can be rewritten as

$$(4^x)^2 - 20 \times 4^x + 84 = 0$$

and hence find the solutions of the equation, giving your answers correct to three decimal places.

b) Show that the equation $2^x + 28 \times (2^{-x}) = 11$ can be rewritten as

$$(2^x)^2 - 11 \times 2^x + 28 = 0$$

and hence find the solutions of the equation, giving your answers correct to three decimal places.

6 If $\log_k 2 = p$ and $\log_k 3 = q$ then we can express $\log_k 18$ in terms of p and q as follows:

$$\begin{aligned}
\log_k 18 &= \log_k(2 \times 3^2) \\
&= \log_k 2 + \log_k(3^2) \\
&= \log_k 2 + 2\log_k 3 \\
&= p + 2q.
\end{aligned}$$

Find expressions for the following in terms of p and q:

a) $\log_k 6$

b) $\log_k 4$

c) $\log_k 12$

d) $\log_k\left(\dfrac{1}{9}\right)$

e) $\log_k\left(\dfrac{2}{3}\right)$

f) $\log_k\left(\dfrac{4}{3}\right)$

g) $\log_k 4.5$

h) $\log_k 54$

7 Solve the inequalities

a) $3.2^x < 5000$

b) $0.7^n > 0.001$

c) $1.3^t > 3$

8 Prove that if $5^{3t-2} = 10^{t+1}$ then $t = \dfrac{\log 250}{\log 12.5}$.

9 a) A man has £3000 invested in an account that pays 6% compound interest per annum. How many years must the money be invested to grow to £10 000?

b) A loaf of bread currently costs 90p. If the rate of inflation is 4.5% each year, in how many years' time will the cost of a loaf be £2.00?

c) Mrs Jones bought a new car for £24 000 on 1st January 2004. The value of the car depreciates at 25% per annum. When will the value of the car be £4500?

10 Which number is bigger, 2004^{2005} or 2005^{2004}?

11 a) Find the value of $\log_{10}(7^{1000})$.
b) How many digits are there in the value of 7^{1000}?
c) What is the last (units) digit of 7^{1000}?

> **Hint:**
> b) If an integer has standard form $a \times 10^n$
>   how many digits does the integer have?
> c) Look for a pattern in the units digits of
> $7^1, 7^2, 7^3, 7^4, 7^5, 7^6, \ldots$

Equations with Logarithms

The approach to equations involving logarithms should be:

- first use the rules of logarithms to collect all the log terms into a single term;
- then use the basic definition of a logarithm to remove the logs altogether;
- solve the resulting equation;
- check that your final answers are sensible for the original equation.

This approach will be illustrated in example 12.

EXAMPLE 12

Solve the equation $\log_4(10x + 6) - 2\log_4 x = 1$.

Start by using the rules of logarithms to collect all the log terms into a single term:

> Using rule 3

$$\log_4(10x + 6) - 2\log_4 x = 1$$

> Using rule 2

$$\Rightarrow \quad \log_4(10x + 6) - \log_4(x^2) = 1$$

$$\Rightarrow \quad \log_4\left(\frac{10x + 6}{x^2}\right) = 1.$$

Now use the basic definition of a logarithm to remove the logs altogether:

$$\Rightarrow \quad \left(\frac{10x + 6}{x^2}\right) = 4^1 = 4.$$

> Remember $\log_4 z = p$
> means that $z = 4^p$.

This equation can now be solved by standard methods:

$$\Rightarrow \quad 10x + 6 = 4x^2$$
$$\Rightarrow \quad 2x^2 - 5x - 3 = 0$$
$$\Rightarrow \quad (2x + 1)(x - 3) = 0$$
$$\Rightarrow \quad x = -\frac{1}{2} \text{ or } 3.$$

Finally check the final answers make sense in the original equation:

if you have $x = -\frac{1}{2}$ then the original equation would require us to evaluate $\log_4\left(-\frac{1}{2}\right)$

but you know that we can only find the logarithm of positive quantities;

if you have $x = 3$ then the original equation requires you to evaluate $\log_4 36$ and $\log_4 2$ and there is no problem with either of these values.

The only root of the equation $\log_4(10x + 6) - 2\log_4 x = 1$ is $x = 3$.

EXAMPLE 13

Solve the simultaneous equations $\left. \begin{array}{l} xy = 108 \\ 2\log_2 x - \log_2 y = 1 \end{array} \right\}$.

Starting with the second equation:

Using rule 3

Using rule 2

$$2\log_2 x - \log_2 y = 1$$
$$\implies \log_2(x^2) - \log_2 y = 1$$
$$\implies \log_2\left(\frac{x^2}{y}\right) = 1$$
$$\implies \frac{x^2}{y} = 2^1 = 2$$

Remember $\log_a z = p$ means that $z = 2^p$.

$$\implies 2y = x^2$$
$$\implies y = \frac{1}{2}x^2.$$

Substituting this into the first equation gives

$$\frac{1}{2}x^3 = 108$$
$$\implies x^3 = 216$$
$$\implies x = 6$$

$$y = \frac{1}{2}x^2 \implies y = 18.$$

EXERCISE 5

These questions require you to take considerable care with your algebra. Remember to check your answers are sensible for the original equation.

1. Solve the equation
   $$\log_4(9x - 2) - 2\log_4 x = 1.$$

2. Solve the equation
   $$\log_5(2x - 3) + \log_5(6x + 1) = 3.$$

3. Solve the equation
   $$\log_6(x + 1) = 3 - \log_6(5x + 11).$$

4. Solve the equation
   $$\log_3(5x + 2) = 1 + 2\log_3(x - 2).$$

5. Solve the simultaneous equations
   $$\begin{cases} x^2 y = 32 \\ \log_2 x + \log_2 y = 4. \end{cases}$$

6. Solve the simultaneous equations
   $$\begin{cases} \log_y x = 2 \\ \log_2 x + \log_2 y = 3. \end{cases}$$

Having studied this chapter you should

● know that if $a > 1$ and $k > 0$ then $y = ka^x$ is an exponential growth curve with graph

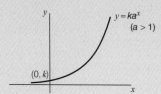

● know that if $0 < a < 1$ and $k > 0$ then $y = ka^x$ is an exponential decay curve with graph

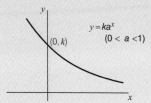

● understand and use the defining property of logarithms:
$x = a^p$ means that $\log_a x = p$

● recognise the graph of $y = \log_a x$ and appreciate that it is the image of $y = a^x$ after a reflection in $y = x$

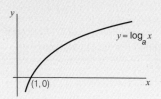

● understand and use the five properties of logarithms:

$\log_a(xy) = \log_a x + \log_a y$

$\log_a\left(\dfrac{x}{y}\right) = \log_a x - \log_a y$

$\log_a(x^n) = n \log_a x$

$\log_a 1 = 0$

$\log_a a = 1$

● be able to use logarithms to solve equations of the form $a^x = b$

REVISION EXERCISE

1   Sketch the graphs of
    a) $y = 3.7^x$ b) $y = 8 \times 0.75^x$ c) $y = \log_7 x$

2   a)   A quantity y grows exponentially. The table shows values of y at different times t.
         Determine the values of a and b.

t	5	10	15	b
y	25	50	a	400

    b)   A quantity z decays exponentially. The table shows values of z at different times t.
         Determine the values of p and q.

t	2	4	6	8
z	1600	1200	p	q

3   Without using a calculator
    a)   determine the value of $\log_9 27$,
    b)   determine the value of $\log_8\left(\dfrac{1}{64}\right)$,
    c)   simplify the expression $\log_{12} 20 + 2 \log_{12} 6 - \log_{12} 5$,
    d)   simplify the expression $2 \log_4 5 - \log_4 50$.

4   Solve, correct to three decimal places, the equations
    a) $8^x = 210$ b) $5^y = 0.081$

    and hence write down the values, correct to three decimal places, of $\log_8 210$ and $\log_5 0.081$

5   Solve the following equations and inequalities:
    a) $5^x = 1000$ b) $3^{2t-1} = 100$
    c) $1.08^y > 3$ d) $0.7^z > 0.001$
    e) $5^{2x} - 11 \times 5^x - 21 = 0$ f) $2^{2x+3} - 18 \times 2^x + 3 = 0$

6   Solve the following equations
    a) $\log_4(7x - 1) - \log_4 5 = 1$ b) $\log_6(2x - 1) + \log_6 4 = 2$

7   Solve the equations
    a) $\log_6(2x - 1) + \log_6(x - 1) = 2$ b) $\log_4(x + 12) - \log_4(x - 18) = 2$
    c) $\log_3(x + 4) - 2 \log_3(x - 2) = -1$

8   A man invests £1000 into a building society account that pays 5% per annum
    compound interest. Find the value of his savings after 3 years.
    Write down an expression for the value of his savings after n years.
    How long does it take for his savings to grow to £2500?

9   Find the smallest integer satisfying the inequality $3^n > 2004^{2005}$.

10 **a)** Explain what is meant by the term "exponential decay". Sketch a graph showing exponential decay.

b) A model giving the number, N, of particles of isotope Z present in a block of radioactive material t hours after the start of an experiment is given by

$$N = 35\,000 \times 0.82^t.$$

i) Find the value of N when $t = 4$.

ii) Find the value of t for which the number of particles of isotope Z present in the block is 10 000.

11 **a)** Prove that if $x > 0$ then $\log_2(x^2 + 4x + 4) - \log_2\left(\dfrac{x+2}{x+3}\right) \equiv \log_2(x^2 + 5x + 6)$.

b) Prove that if $2^{4t+2} = 5^{t+1}$ then

$$t = \log\left(\frac{5}{4}\right)\Big/\log\left(\frac{16}{5}\right).$$

12 A company produces a "high tech" product and monthly sales of this product are currently constant at 40 000 sales per month.

Tomorrow the company will announce the launch of a new product which will eventually replace the original product. The new product will not be available in the shops for six months.

The marketing division estimates that the announcement of the launch will produce a 10% reduction in sales of the original product each month.

a) Write down an expression for the sales of the original product t months after the announcement of the launch of the new product.

b) Sketch a graph to show the marketing division estimates that sales of the original product will progress in the months after the announcement.

c) When will the monthly sales of the original product first fall beneath 1500?

The marketing division estimates that the monthly sales, $S(t)$, of the new product t months after the launch announcement can be modelled by

$$S(t) = \begin{cases} 0 & \text{if} \quad t < 6 \\ 1500 \times 1.15^{t-6} & \text{if} \quad t \geqslant 6 \end{cases}$$

d) Sketch the graph of $S(t)$ and explain its significance. Explain also why it is very unlikely that this equation will model sales for all values of t greater than 6.

e) Find how long it takes for the monthly sales of the new product to exceed 40 000.

f) Find the value of t when the monthly sales of the two products are equal.

13 **a)** Find the exact solutions of the equation $x^3 + 6x^2 - 19x + 6 = 0$.

b) Show that the equation

$$\log_3 x + 2\log_3(x+3) = \log_3(28x - 6)$$

can be rewritten as

$$x^3 + 6x^2 - 19x + 6 = 0.$$

c) Hence solve the equation $\log_3 x + 2\log_3(x+3) = \log_3(28x - 6)$.

7 Arithmetic and Geometric Progressions

The purpose of this chapter is to enable you to

- recognise arithmetic progressions and find the nth term and sum of the first n terms of an arithmetic progression

- recognise geometric progressions and find the nth term and sum of the first n terms of a geometric progression

- find the sum to infinity of a geometric progression whose common ratio lies between -1 and 1

Arithmetic progressions

EXAMPLE 1

On her first birthday, Uncle Nick plans to give his young niece, Saffron, £20 and plans to increase the present each year by £5 until she reaches her 21st birthday.

How much will Saffron receive from Uncle Nick

a) on her 15th birthday?

b) on her 21st birthday?

> Note that the amount for the fifteenth birthday is £20 plus **fourteen** increases each of £5.

a) On her first birthday Saffron will receive   £20

On her second birthday Saffron will receive $20 + 5 = £25$

On her third birthday Saffron will receive $20 + 5 + 5 = £30$

...

On her fifteenth birthday Saffron will receive $20 + 14 \times 5 = £90$.

b) On her twenty first birthday Saffron will receive $20 + 20 \times 5 = £120$.

Sequences such as 20, 25, 30, 35, ... in which each term is obtained by adding a constant number to the previous term are called **arithmetic sequences** or **arithmetic progressions**. The number which is added each time is called the **common difference of the sequence**.

Provided the **first term** and the **common difference** of an arithmetic progression are known then it is possible to write down as many terms of the sequence as you wish.

EXAMPLE 2

An arithmetic progression, $u_1, u_2, u_3, \ldots$ has first term 7 and common difference 2.3.

a) Write down the first five terms of the sequence.

b) Calculate the 80th term of the sequence.

c) Write down an expression for the nth term of the sequence.

EXAMPLE 2 (continued)

a) The first five terms are

$u_1 = 7,$

$u_2 = 7 + 2.3 = 9.3,$

$u_3 = 9.3 + 2.3 = 11.6,$

$u_4 = 11.6 + 2.3 = 13.9,$

$u_5 = 13.9 + 2.3 = 16.2.$

b) $u_{80} = 7 + 79 \times 2.3 = 188.7.$ — Note that the 80th term is 7 plus **79** increases each of 2.3.

c) Similarly,

$u_n = $ nth term $= 7 + (n-1) \times 2.3$ — Note that the nth term is 7 plus $n - 1$ increases each of 2.3.

$= 7 + 2.3n - 2.3$

$= 2.3n + 4.7.$

A more general arithmetic sequence has first term a and common difference d. The first five terms of this sequence are

$a, a + d, a + 2d, a + 3d, a + 4d, \ldots$

The 80th term of this sequence will be

$a + 79d.$

The nth term of this sequence will be

$a + (n-1)d.$

If $u_1, u_2, u_3, \ldots$ is an arithmetic sequence with first term a and common difference d, then

$$u_n = \text{nth term} = a + (n-1)d$$

EXAMPLE 3

How many terms are there in the sequence 3, 4.2, 5.4, ..., 297?

Suppose there are n terms. Then

$297 = \text{nth term} = 3 + (n-1) \times 1.2$

$\Rightarrow \quad 297 = 1.8 + 1.2n$

$\Rightarrow \quad 295.2 = 1.2n$

$\Rightarrow \quad n = 246.$

EXAMPLE 4

An arithmetic sequence has third term 37 and eighth term 29.
Find the first term and common difference of the sequence and hence find how many terms of this sequence are positive.

Suppose the first term is a and the common difference is d.

Using the information given about the third and eighth terms, a pair of simultaneous equations for a and d can be obtained:

$$\left.\begin{array}{l} 37 = \text{3rd term} = a + 2d \\ 29 = \text{8th term} = a + 7d \end{array}\right\} \Rightarrow d = -1.6,\ a = 40.2$$

$$\begin{aligned} n\text{th term} &= 40.2 + (n-1) \times (-1.6) \\ &= 40.2 - 1.6n + 1.6 \\ &= 41.8 - 1.6n. \end{aligned}$$

You require the nth term of the sequence to be positive so

$$41.8 - 1.6n > 0$$
$$\Rightarrow \quad 41.8 > 1.6n$$
$$\Rightarrow \quad n < \frac{41.8}{1.6}$$
$$\Rightarrow \quad n < 26.125 \qquad \text{so 26 terms of the sequence are positive.}$$

EXERCISE 1

1. Find the 10th, 31st and 59th terms of the sequence 4, 5.3, 6.6, ...
   Find also an expression for the nth term of this sequence.

2. Find the 10th, 31st and 59th terms of the sequence 12, 7, 2, –3, ...
   Find also an expression for the nth term of this sequence.

3. Find expressions for the nth terms of the sequences
   a) 2, 5, 8, 11, ...
   b) 3.2, 3.5, 3.8, 4.1, ...
   c) 103, 101, 99, 97, ...
   d) –12, –5, 2, 9, ...

4. How many terms are there in these arithmetic sequences?
   a) 2, 5, 8, 11, ..., 3782
   b) 124, 119, 114, ..., –71
   c) 3.7, 5, 6.3, ..., 330

5. How many terms in the arithmetic sequence 5, 7.1, 9.2, 11.3, ... are less than 800?

6. How many positive terms are there in the arithmetic sequence

   121.3, 119.5, 117.7, 115.9, ...?

7. An arithmetic sequence has sixth term 21 and tenth term 27. Find the first term of the sequence and the common difference. Hence find the number of terms in the sequence that are less than 120.

8. A triangle has angles which form an arithmetic progression. The largest angle is four times the smallest angle. Find the sizes of the three angles.

9 An arithmetic progression $u_1, u_2, u_3, \ldots$ has $u_5 = 80$ and $u_{12} = 57.6$.
How many positive terms does the progression have?

10 An arithmetic progression $v_1, v_2, v_3, \ldots$ has first term 12 and common difference 1.6.
How many terms of the progression are greater than 1000 but smaller than 2000?

The Sum of an Arithmetic Progression

EXAMPLE 5

On her first birthday, Uncle Nick plans to give his young niece, Saffron, £20 and plans to increase the present each year by £5 until she reaches her 21st birthday.
How much will Saffron have received altogether from Uncle Nick by the time she has celebrated her 21st birthday?

If A_i is the amount, in pounds, that Saffron receives from Uncle Nick on her ith birthday then $A_1, A_2, A_3 \ldots$ is an arithmetic progression with first term 20 and common difference 5.

The amount she receives on her 21st birthday will be A_{21} where

$$A_{21} = 20 + 20 \times 5 = £120.$$

Let the total amount, in pounds, that she will have received by her 21st birthday be S_{21}.
You know that

S_{21}	=	20	+	25	+	30	+	$\cdots$	+	110	+	115	+	120

This sum can be evaluated easily using a neat trick: simply write out the sum again but start with the final term.

S_{21}	=	20	+	25	+	30	+	$\cdots$	+	110	+	115	+	120
S_{21}	=	120	+	115	+	110	+	$\cdots$	+	30	+	25	+	20

and then add the two rows, column by column:

S_{21}	=	20	+	25	+	30	+	$\cdots$	+	110	+	115	+	120
S_{21}	=	120	+	115	+	110	+	$\cdots$	+	30	+	25	+	20
$2S_{21}$	=	140	+	140	+	140	+	$\cdots$	+	140	+	140	+	140

You then obtain

$$2S_{21} = 21 \times 140$$

$$\Rightarrow \quad S_{21} = \frac{21 \times 140}{2} = 1470$$

so Saffron will have received £1470 from Uncle Nick by the time she has celebrated her 21st birthday.

The method used in this example to find the total of the presents that Uncle Nick will have given Saffron after her 21st birthday can be generalised to find the sum of any arithmetic progression.

Suppose you wish to find the sum, S_n, of the first n terms of the arithmetic progression

$$a, a+d, a+2d, a+3d, a+4d \ldots$$

The last term, L, of the sum is given by

$$L = n\text{th term} = a + (n-1)d.$$

The penultimate term will be $L - d$ and the term before that will be $L - 2d$
so

$$S_n \quad = \quad a \quad + \quad a+d \quad + \quad a+2d \quad + \quad \cdots \quad + \quad L-2d \quad + \quad L-d \quad + \quad L$$

Now write the sum out again in a second row, but this time start with the final term:

$$\begin{aligned}
S_n &= a &+& a+d &+& a+2d &+& \cdots &+& L-2d &+& L-d &+& L \\
S_n &= L &+& L-d &+& L-2d &+& \cdots &+& a+2d &+& a+d &+& a
\end{aligned}$$

and add the two rows, column by column to obtain

$$\begin{aligned}
S_n &= a &+& a+d &+& a+2d &+& \cdots &+& L-2d &+& L-d &+& L \\
S_n &= L &+& L-d &+& L-2d &+& \cdots &+& a+2d &+& a+d &+& a \\
\hline
2S_n &= a+L &+& a+L &+& a+L &+& \cdots &+& a+L &+& a+L &+& a+L
\end{aligned}$$

From this, it can be seen that

> Each term in the original sequence contributes one $(a+L)$ to the final row. Since there are n terms in the sequence, there will be n lots of $(a+L)$ on the bottom row.

$$2S_n = n \times (a+L)$$

$$\Rightarrow \quad S_n = \frac{1}{2}n(a+L).$$

Since

$$L = n\text{th term} = a + (n-1)d,$$

the formula $\quad S_n = \frac{1}{2}n(a+L)$

can be rewritten as

$$S_n = \frac{1}{2}n(a + (a + (n-1)d))$$

$$\Rightarrow \quad S_n = \frac{1}{2}n(2a + (n-1)d).$$

If S_n is the sum of the first n terms of an arithmetic progression with first term a and common difference d then

$$S_n = \frac{1}{2}n(2a + (n-1)d) = \frac{1}{2}n(a+L)$$

where L denotes the nth term.

EXAMPLE 6

Find the sum of the following arithmetic progression:

$$1 + 3 + 5 + 7 + \cdots + 999.$$

S O L U T I O N

The arithmetic progression has first term 1 and common difference 2.

To find the sum you need to know the number of terms in the progression. Suppose there are n terms then

$$999 = n\text{th term} = 1 + (n - 1) \times 2$$
$$\Rightarrow \quad 999 = 2n - 1$$
$$\Rightarrow \quad 1000 = 2n$$
$$\Rightarrow \quad n = 500.$$

You now want the sum of the arithmetic progression of 500 terms with first term 1 and last term 999, so

$$S_{500} = \frac{1}{2} n(a + L) = \frac{1}{2} \times 500 \times (1 + 999) = 250\,000.$$

EXAMPLE 7

An arithmetic progression has first term 5 and seventh term 8.

a) Find the sum of the first 60 terms of the progression.
b) How many terms of the progression must be added together to obtain a total that exceeds 100 000?

S O L U T I O N

a) You must first find the common difference of the progression:

$$7\text{th term} = a + 6d$$
$$\Rightarrow \quad 8 = 5 + 6d$$
$$\Rightarrow \quad d = \frac{1}{2}$$

and you can now calculate the sum of the first 60 terms:

$$S_n = \frac{1}{2} n(2a + (n - 1)d)$$
$$\Rightarrow \quad S_{60} = \frac{1}{2} \times 60\left(10 + 59 \times \frac{1}{2}\right) = 1185.$$

EXAMPLE 7 (continued)

b) Suppose that $S_n > 100\,000$

$$\Rightarrow \quad \frac{1}{2}n\left(10 + (n-1) \times \frac{1}{2}\right) > 100\,000$$

$$\Rightarrow \quad \frac{1}{2}n\left(10 + \frac{1}{2}n - \frac{1}{2}\right) > 100\,000$$

$$\Rightarrow \quad \frac{1}{2}n\left(\frac{19}{2} + \frac{1}{2}n\right) > 100\,000$$

$$\Rightarrow \quad \frac{1}{4}n^2 + \frac{19}{4}n > 100\,000$$

$$\Rightarrow \quad n^2 + 19n > 400\,000$$

$$\Rightarrow \quad n^2 + 19n - 400\,000 > 0$$

$$\Rightarrow \quad n > 623.02\ldots$$

$$\text{or} \quad n < -642.02\ldots$$

Using a graphical calculator to solve the quadratic equation $n^2 + 19n - 400\,000 = 0$ gives $n = 623.02\ldots$ or $-642.02\ldots$

The graph of $y = n^2 + 19n - 400\,000$ will be a $\cup$ shaped parabola.

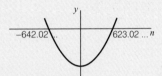

n must be positive so you can ignore the negative solutions

$$\Rightarrow \quad n > 623.02\ldots$$

The solution of the inequality is therefore $n > 623.02\ldots$ or $n < 642.02\ldots$

n must also be an integer so you can conclude that at least 624 terms of the sequence must be added together to get a sum that exceeds 100 000.

The Sum of the First n Natural Numbers

Remember that the **natural numbers** are just the basic counting numbers: 1, 2, 3, 4, ...

Using the result for the sum of an arithmetic progression, you can quickly calculate the sum of the first 600 natural numbers:

$$1 + 2 + 3 + \cdots + 599 + 600 = \frac{1}{2} \times 600 \times (1 + 600) = 300 \times 601 = 180\,300.$$

Similarly, the sum of the first n natural numbers can be calculated as

$$1 + 2 + 3 + \cdots + (n-1) + n = \frac{1}{2} \times n \times (1 + n) = \frac{1}{2}n(n+1).$$

So you have the important result

$$\sum_{r=1}^{n} r = 1 + 2 + 3 + \cdots + n = \frac{1}{2}n(n+1)$$

EXAMPLE 8

Without using a calculator, find the value of $\sum_{1}^{100} (5r - 2)$.

Solution 1: You start by writing the sum out longhand:

$$\sum_{1}^{100} (5r + 2) = 7 + 12 + 17 + 22 + 27 + \cdots + 502$$

and this is the sum of the 100 terms of an arithmetic progression with first term 7 and last term 502, so

> The first term ($r = 1$) is $5 \times 1 + 2 = 7$.
> The second term ($r = 2$) is $5 \times 2 + 2 = 12$.
> The third term ($r = 3$) is $5 \times 3 + 2 = 17$.
> ...
> The 100th term ($r = 100$) is $5 \times 100 + 2 = 502$.

$$\sum_{1}^{100} (5r - 2) = \frac{1}{2}n(a + L)$$

$$= \frac{1}{2} \times 100 \times (7 + 502)$$

$$= \frac{1}{2} \times 50\,900$$

$$= 25\,450.$$

Solution 2: Recalling the properties of sigma notation introduced in chapter 4:

$$\sum_{r=1}^{n} (a_r + b_r) = \sum_{r=1}^{n} a_r + \sum_{r=1}^{n} b_r$$

$$\sum_{r=1}^{n} c a_r = c \sum_{r=1}^{n} a_r$$

$$\sum_{r=1}^{n} c = cn.$$

You can write

$$\sum_{1}^{100} (5r + 2) = \sum_{r=1}^{100} 5r + \sum_{r=1}^{100} 2$$

$$= 5 \sum_{r=1}^{100} r + \sum_{r=1}^{100} 2$$

$$= 5 \times \frac{1}{2} \times 100 \times 101 + 2 \times 100$$

$$= \frac{1}{2} \times 50\,500 + 200$$

$$= 25\,250 + 200$$

$$= 25\,450.$$

> You know that
> $$\sum_{r=1}^{100} r = \frac{1}{2} \times 100 \times (100 + 1)$$
> and that
> $$\sum_{r=1}^{100} 2 = 2 \times 100$$

EXERCISE 2

1 Find the sum of the following arithmetic progressions:
   a) $5.2 + 6.4 + 7.6 + \cdots + 245.2$
   b) $3 + 7 + 11 + 15 + \cdots$ (72 terms in the sum).

2 An arithmetic progression has 51 terms. Its first term is 3 and the sum of the progression is 280.5. Find
   a) the last term of the progression;   **b)** the common difference of the progression;
   c) the fifth term of the progression.

3 12 metres of insulating tape are wound onto a reel of circumference 12 cm. Due to the thickness of the tape, each turn is 1.8 mm longer than the previous one. How many complete turns are needed to wind the tape onto the reel?

4 The sixth term of an arithmetic progression is twice the third term and the first term is 3. Calculate
   a) the common difference,   **b)** the tenth term,
   c) the sum of the first 20 terms.

5 **a)** Find the sum of the first 250 positive integers.
   b) Find the value of $\displaystyle\sum_{10}^{50} r$.
   c) Find the value of k if $\displaystyle\sum_{r=10}^{k} r = 516$.

6 How many terms of the arithmetic progression $5 + 7 + 9 + \cdots$ are required to make a sum of 1517?

7 How many terms of the arithmetic progression $6 + 13 + 20 + 27 + \cdots$ are required to get a sum that exceeds 5000?

8 The smallest four-digit number which is exactly divisible by 9 is 1008. What is the largest such number? How many four-digit numbers are divisible by 9? What is the sum of these numbers?

9 The sum of the first n terms of the arithmetic progression

   $13 + 16.5 + 20 + \cdots$

is the same as the sum of the first n terms of the arithmetic progression

   $3 + 7 + 11 + \cdots$.

Calculate the value of n.

10 **a)** Find the sum of all the positive integers less than 1000.
   b) Find the sum of all the positive integers less than 1000 that are divisible by 7.
   c) Find the sum of all the positive integers less than 1000 that are not divisible by 7.
   d) Find the sum of all the positive integers less than 1000 that are divisible by 5 or by 7.

11 **a)** Evaluate $\displaystyle\sum_{r=1}^{20} (3r - 2)$.

   b) Find the smallest integer n such that $\displaystyle\sum_{r=1}^{n} (3r - 2) > 5000$.

Geometric Progressions

EXAMPLE 9

On her first birthday, Uncle Michael plans to give his young niece, Saffron, £20 and plans to increase the present each year by 10% until she reaches her 21st birthday.

How much will Saffron receive from Uncle Michael

> Each year the present increases by 10% so the new present is 1.1 times the previous present.

a) on her 15th birthday?
b) on her 21st birthday?

S O L U T I O N

a) On her first birthday Saffron will receive   £20
On her second birthday Saffron will receive $20 \times 1.1 = £22$
On her third birthday Saffron will receive $20 \times 1.1 \times 1.1 = £24.20$
...
On her fifteenth birthday Saffron will receive $20 \times 1.1^{14} = £75.95$
(to the nearest penny).

> Note that the amount for the fifteenth birthday is £20 times 1.1^{14} because there have been **fourteen** increases each of 10%.

b) Similarly, on her 21st birthday Saffron will receive $20 \times 1.1^{20} = £134.55$
(to the nearest penny).

Sequences such as

$20, 22, 24.20, 26.62, ...$ $1, 2, 4, 8, 16, ...$ $3, -6, 12, -24, 48, ...$
$3, 12, 48, 192, ...$ $27, 9, 3, 1, \frac{1}{3}, ...$

in which each term is obtained from the previous term by multiplying by a constant number are called **geometric progressions**. The constant number which we multiply by is called the **common ratio**.

EXAMPLE 10

A geometric progression $u_1, u_2, u_3, ...$ has first term 7 and common ratio 2.

a) Write down the first five terms of the sequence.
b) Calculate the 10th term of the sequence.
c) Write down an expression for the nth term of the sequence.

S O L U T I O N

a) The first five terms are

$u_1 = 7,$
$u_2 = 7 \times 2 = 14,$
$u_3 = 14 \times 2 = 28,$
$u_4 = 28 \times 2 = 56,$
$u_5 = 56 \times 2 = 112.$

> You have to multiply through by 2 **nine** times to reach the tenth term.

b) $u_{10} = 7 \times 2^9 = 3584.$

c) Similarly, $u_n = n$th term $= 7 \times 2^{n-1}.$

> You have to multiply through by 2 a total of $n - 1$ times to reach the nth term.

A more general geometric progression has first term a and common ratio r.
The first five terms of this sequence are

$a, ar, ar^2, ar^3, ar^4, \ldots$

The 80th term of this sequence will be ar^{79} and the nth term of this sequence will be ar^{n-1}.

> If $u_1, u_2, u_3 \ldots$, is a geometric progression with first term a and common ratio r,
>
> $u_n = n$th term $= ar^{n-1}$

EXAMPLE 11

How many terms in the geometric progression

$10, 12, 14.4, 17.28, \ldots$

are less than 1000?

The geometric progression has first term 10 and common ratio 1.2 so the nth term is $10 \times 1.2^{n-1}$.

Suppose the nth term is less than 1000. Then

$$10 \times 1.2^{n-1} < 1000$$

> This is very similar to the inequalities considered in chapter 6: it can be solved by finding the log of each side.

$$\Rightarrow \quad 1.2^{n-1} < 100$$

$$\Rightarrow \quad \log(1.2^{n-1}) < \log 100$$

> Remember that $\log(x^p) = p \log x$ and that $\log 100 = 2$ since $100 = 10^2$.

$$\Rightarrow \quad (n-1)\log 1.2 < 2$$

$$\Rightarrow \quad n - 1 < \frac{2}{\log 1.2}$$

> $\log 1.2 = 0.07818 \ldots$ which is positive.
>
> You can therefore divide each side of the inequality by $\log 1.2$ without needing to reverse the inequality.

$$\Rightarrow \quad n - 1 < 25.258 \ldots$$

$$\Rightarrow \quad n < 26.258 \ldots$$

So 26 terms of the progression are less than 1000.

> As a check:
> 26th term $= 10 \times 1.2^{25} = 953.96 \ldots$
> but
> 27th term $= 10 \times 1.2^{26} = 1144.75 \ldots$

EXAMPLE 12

A geometric sequence has third term 288 and sixth term 121.5. Determine the first term and common ratio of the sequence and hence find the number of terms that are greater than 1.

If the first term is a and the common ratio is r then

third term $= ar^2 = 288$ [1]

sixth term $= ar^5 = 121.5$ [2]

EXAMPLE 12 (continued)

Dividing equation [2] by equation [1] gives $\dfrac{ar^5}{ar^2} = \dfrac{121.5}{288} = \dfrac{1215}{2880} = \dfrac{27}{64}$

$$\Rightarrow \quad r^3 = \frac{27}{64}$$

$$\Rightarrow \quad r = \frac{3}{4}.$$

Equation [1] now gives

$$a \times \frac{9}{16} = 288$$

$$\Rightarrow \quad a = 512.$$

The nth term of the sequence with first term 512 and common ratio $\dfrac{3}{4}$ is $512 \times \left(\dfrac{3}{4}\right)^{n-1}$.

If the nth term is greater than 1 then

$$512 \times \left(\frac{3}{4}\right)^{n-1} > 1$$

$$\Rightarrow \quad \left(\frac{3}{4}\right)^{n-1} > \frac{1}{512}$$

$$\Rightarrow \quad \log\left[\left(\frac{3}{4}\right)^{n-1}\right] > \log\left(\frac{1}{512}\right)$$

$$\Rightarrow \quad (n-1)\log\left(\frac{3}{4}\right) > \log\left(\frac{1}{512}\right)$$

$\log\left(\dfrac{3}{4}\right) = -0.1249 \ldots$ which is negative. Dividing each side of the inequality by $\log\left(\dfrac{3}{4}\right)$ will **reverse the direction of the inequality.**

$$\Rightarrow \quad n-1 < \frac{\log\left(\dfrac{1}{512}\right)}{\log\left(\dfrac{3}{4}\right)} = 21.6847 \ldots$$

$$\Rightarrow \quad n < 22.6847 \ldots.$$

So 22 terms of the sequence are greater than 1.

As a check:
22nd term $= 512 \times 0.75^{21} = 1.217 \ldots$
but
23rd term $= 512 \times 0.75^{22} = 0.913 \ldots$

EXAMPLE 13

The ages of three sisters are different and are the first three terms of a geometric progression and are also the first, third and sixth terms of an arithmetic progression. Find the common ratio of the geometric progression.
If the total age of the three sisters is 57 years, find the age of each of the sisters.

EXAMPLE 13 (continued)

Let a and r be the first term and the common ratio of the geometric progression.

The ages of the three sisters are a, ar and ar^2.

If d is the common difference of the arithmetic progression then

$$ar = \text{age of second sister} = \text{third term of arithmetic progression} = a + 2d$$

$$\Rightarrow \quad d = \frac{ar - a}{2}.$$

Similarly

$$ar^2 = \text{age of second sister} = \text{sixth term of arithmetic progression} = a + 5d$$

$$\Rightarrow \quad d = \frac{ar^2 - a}{5}.$$

Combining the two equations gives $\dfrac{ar - a}{2} = \dfrac{ar^2 - a}{5}$

$$\Rightarrow \quad 5ar - 5a = 2ar^2 - 2a$$

$$\Rightarrow \quad 0 = 2ar^2 - 5ar + 3a$$

$$\Rightarrow \quad 0 = a(2r^2 - 5r + 3)$$

$$\Rightarrow \quad 0 = a(r - 1)(2r - 3)$$

$$\Rightarrow \quad a = 0 \quad \text{or} \quad r = 1 \quad \text{or} \quad r = \frac{3}{2}.$$

$a = 0$ would mean all three sisters were 0 years old, but you know that the ages of the sisters are different; $r = 1$ would mean all three sisters were a years old but, again, you know that the ages of the sisters are different

$$\Rightarrow \quad r = \frac{3}{2}.$$

The ages of the sisters are a, $\dfrac{3}{2}a$ and $\dfrac{9}{4}a$ so their total age is $\dfrac{19}{4}a$.

$$\frac{19}{4}a = 57 \quad \Rightarrow \quad a = 12.$$

The sisters are 12, 18 and 27.

EXERCISE 3

1 Find the 9th, 33rd and nth terms of the sequences

a) 3, 12, 48, 192, ...   **b)** 27, 9, 3, 1, $\dfrac{1}{3}$, ...   **c)** 3, −6, 12, −24, 48, ...

2 Find the value of k if 1 062 882 is the kth term of the geometric progression

2, 6, 18, 54, ...

3 Find the number of terms in the geometric progressions

a) $3, 12, 48, \ldots, 786\,432$ b) $256, 128, 64, \ldots, \dfrac{1}{1024}$.

4 How many terms are there in the geometric progression

$3, 15, 75, 375, \ldots$

which are less than 1 000 000?

5 A geometric progression has sixth term 9.6 and ninth term 76.8. Find the first term and common ratio of the progression and hence find the thirteenth term of the progression.

6 A geometric progression of positive numbers has second term 12 and sixth term 2. Find, correct to three decimal places, the twelfth term of the progression.

7 How many terms in the geometric progression

$2000, 1600, 1280, 1024 \ldots$

are greater than 2?

8 The first three terms of a geometric progression are the first, third and thirteenth terms of an arithmetic progression.
a) Find the common ratio of the geometric progression.

The third term of the geometric progression is 75.
b) Find the first five terms of the arithmetic progression.

The Sum of a Geometric Progression

EXAMPLE 14

On her first birthday, Uncle Michael plans to give his young niece, Saffron, £20 and plans to increase the present each year by 10% until she reaches her 21st birthday.

How much will Saffron have received altogether from Uncle Michael by the time she has celebrated her 21st birthday?

The amount that Saffron receives on her 21st birthday will be 20×1.1^{20}.
Let the total amount, in pounds, that she will have received by her 21st birthday be S_{21}.

You know that

$$S_{21} = 20 + 20 \times 1.1 + 20 \times 1.1^2 + 20 \times 1.1^3 + \cdots 20 \times 1.1^{19} + 20 \times 1.1^{20}$$

This sum can also be evaluated using a simple trick: multiply the sum through by 1.1 and write down the original sum underneath.

$$1.1S_{21} = \qquad 20 \times 1.1 + 20 \times 1.1^2 + 20 \times 1.1^3 + \cdots 20 \times 1.1^{19} + 20 \times 1.1^{20} + 20 \times 1.1^{21}$$
$$S_{21} = 20 + 20 \times 1.1 + 20 \times 1.1^2 + 20 \times 1.1^3 + \cdots 20 \times 1.1^{19} + 20 \times 1.1^{20}$$

Then subtract the second row from the first row to obtain

$$1.1S_{21} = \qquad 20 \times 1.1 + 20 \times 1.1^2 + 20 \times 1.1^3 + \cdots 20 \times 1.1^{19} + 20 \times 1.1^{20} + 20 \times 1.1^{21}$$
$$S_{21} = 20 + 20 \times 1.1 + 20 \times 1.1^2 + 20 \times 1.1^3 + \cdots 20 \times 1.1^{19} + 20 \times 1.1^{20}$$
$$\ominus \overline{\qquad\qquad\qquad\qquad\qquad\qquad\qquad\qquad\qquad\qquad\qquad\qquad\qquad\qquad\qquad\qquad\qquad}$$
$$0.1S_{21} = -20 + \quad 0 \quad + \quad 0 \quad + \quad 0 \quad + \cdots \quad 0 \quad + \quad 0 \quad + 20 \times 1.1^{21}$$

EXAMPLE 14 (continued)

So $\quad 0.1 S_{21} = 20 \times 1.1^{21} - 20$

$\Rightarrow \quad S_{21} = \dfrac{20 \times 1.1^{21} - 20}{0.1} = 1280.049 \ldots$

and Saffron will have received £1280 (to the nearest pound) by the time she has celebrated her 21st birthday.

Suppose you wish to find the sum, S_n, of the first n terms of the geometric progression

$\quad a, ar, ar^2, ar^3, ar^4, \ldots$

You know that the last term, L, of the sum is given by

$\quad L = n\text{th term} = ar^{n-1}.$

You know that

$S_n \quad = \quad a \quad + \quad ar \quad + \quad ar^2 \quad + \quad ar^3 \quad + \quad \cdots \quad ar^{n-2} \quad + \quad ar^{n-1}$

This sum can also be evaluated by multiplying the sum through by r and writing down the original sum underneath.

$rS_n \quad = \qquad\qquad ar \quad + \quad ar^2 \quad + \quad ar^3 \quad + \quad \cdots \quad ar^{n-2} \quad + \quad ar^{n-1} \quad + \quad ar^n$

$S_n \quad = \quad a \quad + \quad ar \quad + \quad ar^2 \quad + \quad ar^3 \quad + \quad \cdots \quad ar^{n-2} \quad + \quad ar^{n-1}$

The second row must then be subtracted from the first row to obtain

$rS_n \quad = \qquad\qquad ar \quad + \quad ar^2 \quad + \quad ar^3 \quad + \quad \cdots \quad ar^{n-2} \quad + \quad ar^{n-1} \quad + \quad ar^n$

$\ominus \quad S_n \quad = \quad a \quad + \quad ar \quad + \quad ar^2 \quad + \quad ar^3 \quad + \quad \cdots \quad ar^{n-2} \quad + \quad ar^{n-1}$

$\overline{(r-1)S_n = \quad -a \quad + \quad 0 \quad + \quad 0 \quad + \quad 0 \quad + \quad \cdots \quad 0 \quad + \quad 0 \quad + \quad ar^n}$

So $\qquad (r-1)S_n = ar^n - a$

$\Rightarrow \qquad S_n = \dfrac{ar^n - a}{r - 1} = \dfrac{a(r^n - 1)}{r - 1}.$

In cases where the common ratio is smaller than 1 it is more convenient to use an equivalent formula, obtained by multiplying top and bottom of the fraction by -1:

$\quad S_n = \dfrac{a(1 - r^n)}{1 - r}.$

If S_n is the sum of the first n terms of a geometric progression with first term a and common ratio r then

$\quad S_n = \dfrac{a(r^n - 1)}{r - 1} = \dfrac{a(1 - r^n)}{1 - r}.$

The formula book gives the result as

$\quad S_n = \dfrac{a(1 - r^n)}{1 - r}$

but it is worth knowing both forms of the result.

EXAMPLE 15

Find the value of the sum of the first 25 terms of the geometric progression

| 20 | 24 | 28.8 | 34.56 | ... |

The progression has first term 20 and common ratio 1.2.

$$S_{25} = \frac{a(r^n - 1)}{r - 1}$$

$$= \frac{20(1.2^{25} - 1)}{1.2 - 1}$$

$$= 9439.62 \ldots$$

$$= 9440 \quad \text{(correct to 4 s.f.).}$$

EXAMPLE 16

A geometric progression of positive terms has third term 128 and fifth term 327.68.

a) Find the first term and common ratio of the progression.
b) Find the sum of the first 20 terms of this sequence.
c) How many terms of the progression must be added together to obtain a total exceeding 10^8?

a) Third term $= ar^2 = 128$.
Fifth term $= ar^4 = 327.68$.
Dividing the second equation by the first gives

$$\frac{ar^4}{ar^2} = \frac{327.68}{128}$$

$$\Rightarrow \quad r^2 = 2.56$$

$$\Rightarrow \quad r = \pm 1.6.$$

All the terms of the progression must be positive, so a and r must both be positive

$$\Rightarrow \quad r = 1.6.$$

You know that $ar^2 = 128$ so $a = \frac{128}{1.6^2} = 50$.

b) Using the result $S_n = \frac{a(r^n - 1)}{r - 1}$ gives

$$S_{20} = \frac{50(1.6^{20} - 1)}{1.6 - 1} = 1\ 007\ 354.85 \ldots$$

so $S_{20} = 1\ 007\ 000$ to four significant figures.

EXAMPLE 16 (continued)

c) If $S_n > 10^8$ then

$$\frac{50(1.6^n - 1)}{1.6 - 1} > 10^8$$

$$\Rightarrow \quad 50(1.6^n - 1) > 6 \times 10^7$$

$$\Rightarrow \quad 1.6^n - 1 > 1.2 \times 10^6$$

$$\Rightarrow \quad 1.6^n > 1\,200\,001$$

$$\Rightarrow \quad \log(1.6^n) > \log 1\,200\,001$$

$$\Rightarrow \quad n \log 1.6 > \log 1\,200\,001$$

$$\Rightarrow \quad n > \frac{\log 1\,200\,001}{\log 1.6}$$

$$\Rightarrow \quad n > 29.78 \ldots$$

so 30 terms must be added together to obtain a sum greater than 10^8.

EXAMPLE 17

A sequence $u_1, u_2, u_3, \ldots$ is defined by

$$u_1 = 60, \quad u_{n+1} = 0.8u_n.$$

Find the value of $\sum_{r=1}^{20} u_r$.

Since each term in the sequence is 0.8 times the previous term, the sequence is a geometric progression with common ratio 0.8. The first term is 60.

$$\sum_{r=1}^{20} u_r = u_1 + u_2 + \cdots + u_{20} = S_{20} = \frac{60(1 - 0.8^{20})}{1 - 0.8} = 296.5 \qquad \text{(to 4 s.f.)}.$$

The Sum to Infinity of a Geometric Progression

Now consider the geometric progression $5 + \dfrac{10}{3} + \dfrac{20}{9} + \dfrac{40}{27} + \dfrac{80}{81} + \cdots$.

The progression has first term 5 and common ratio $\dfrac{2}{3}$.

If the sum of the first n terms is S_n then

$$S_n = \frac{5\left(1 - \left(\frac{2}{3}\right)^n\right)}{1 - \frac{2}{3}} = \frac{5\left(1 - \left(\frac{2}{3}\right)^n\right)}{\frac{1}{3}} = 15\left(1 - \left(\frac{2}{3}\right)^n\right).$$

As n gets larger and larger, $\left(\dfrac{2}{3}\right)^n \to 0$ so the value of S_n gets closer and closer to 15.

It can be said that 15 is the **sum to infinity** of this geometric progression.

You can illustrate this arithmetically in a table and geometrically on a number line.

n	S_n
1	5
2	8.33333333
3	10.5555556
4	12.037037
5	13.0246914
6	13.6831276
7	14.122085
8	14.4147234
9	14.6098156
10	14.7398771
20	14.9954891
30	14.9999218

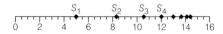

In both the table and the diagram it can be seen that, as the value of n increases, the value of S_n gets closer and closer to 15.

In general, the sum, S_n, of the first n terms of the geometric progression

$$a + ar + ar^2 + ar^3 + ar^4 + \cdots$$

is given by $S_n = \dfrac{a(1 - r^n)}{1 - r}$.

If $-1 < r < 1$ then $r^n \to 0$ as $n \to \infty$, which means that $S_n \to \dfrac{a}{1 - r}$ as $n \to \infty$ and we say that $\dfrac{a}{1 - r}$ is the sum to infinity of the progression.

If $-1 < r < 1$ then the geometric progression

$$a + ar + ar^2 + ar^3 + ar^4 + \cdots$$

has a sum to infinity of

$$\frac{a}{1 - r}.$$

EXAMPLE 18

Find the sums to infinity of the two geometric progressions with first term 5 and third term 2.45, giving your answers exactly as fractions.

If a geometric progression has first term 5 and common ratio is r then

$$\text{third term} = 5r^2$$
$$\Rightarrow \quad 5r^2 = 2.45$$
$$\Rightarrow \quad r^2 = 0.49$$
$$\Rightarrow \quad r = \pm 0.7.$$

In both cases the sum to infinity of the geometric progressions will exist since the common ratio lies between -1 and 1.

If $r = 0.7$ then $\qquad$ sum to infinity $= \dfrac{a}{1 - r} = \dfrac{5}{1 - 0.7} = \dfrac{5}{0.3} = \dfrac{50}{3}.$

If $r = -0.7$ then $\qquad$ sum to infinity $= \dfrac{a}{1 - r} = \dfrac{5}{1 - (-0.7)} = \dfrac{5}{1.3} = \dfrac{50}{13}.$

EXAMPLE 19

Find the first four terms of the geometric progressions which have a sum to infinity of 64 and a second term of 12.

If the first term of the geometric progression is a and the common ratio is r then

sum to infinity $= \dfrac{a}{1-r} = 64$

second term $= ar = 12$

$$\dfrac{a}{1-r} = 64 \quad \Rightarrow \quad a = 64(1-r)$$

$$ar = 12 \quad \Rightarrow \quad 64(1-r)r = 12$$
$$\Rightarrow \quad 16r(1-r) = 3$$
$$\Rightarrow \quad 16r - 16r^2 = 3$$
$$\Rightarrow \quad 16r^2 - 16r + 3 = 0$$
$$\Rightarrow \quad (4r-1)(4r-3) = 0$$
$$\Rightarrow \quad r = \dfrac{1}{4} \quad \text{or} \quad \dfrac{3}{4}.$$

If $r = \dfrac{1}{4}$ then $a = 64(1-r) = 64\left(1 - \dfrac{1}{4}\right) = 48$ so the first four terms of the progression are

48,   12,   3,   0.75.

If $r = \dfrac{3}{4}$ then $a = 64(1-r) = 64\left(1 - \dfrac{3}{4}\right) = 16$ so the first four terms of the progression are

16,   12,   9,   6.75.

EXERCISE 4

1 Find the sums of the following geometric progressions:
  a) $4 + 12 + 36 + \cdots$ (17 terms)
  b) $8 + 12 + 18 + \cdots$ (10 terms)
  c) $5 - 10 + 20 - 40 + \cdots$ (13 terms)
  d) $5 - 10 + 20 - 40 + \cdots$ (14 terms)
  e) $2 + 10 + 50 + \cdots + 781\ 250$
  f) $6 + 3 + 1.5 + \cdots$ (sum to infinity)
  g) $0.7 + 0.07 + 0.007 + 0.0007 + \cdots$ (sum to infinity)

2 The sixth term of a GP is 10 and the eleventh term is 320. Find the first term, the common ratio and the sum of the first 11 terms.

3 A GP is such that the sum of the first eight terms is five times the sum of the first four terms. Determine the possible values of the common ratio.

4 How many terms of the geometric progression $8 + 20 + 50 + \cdots$ must be taken to get a sum that is greater than one hundred million?

5 A company's annual turnover is expected to be 50 million pounds next year and in following years is expected to increase at a rate of 4% per annum.
   a) What will be the company's expected turnover in ten years' time?
   b) After how many years will the company's expected annual turnover first exceed 200 million pounds?
   c) What is the company's total expected turnover for the next ten years?
   d) After how many years will the company's total expected turnover first exceed 2 billion pounds?

6 a) Find the sum to infinity of the geometric progression

   $$30 + 21 + 14.7 + 10.29 + \cdots$$

   b) How many terms of the progression $30 + 21 + 14.7 + 10.29 + \cdots$ must be taken to obtain a sum that is within 10^{-6} of the progression's sum to infinity.

7 a) Prove that $r = \dfrac{1}{2}$ is a root of the equation $16r^3 - 16r^2 + 2 = 0$ and hence solve the equation $16r^3 - 16r^2 + 2 = 0$ completely.
   b) A geometric progression has third term 2 and sum to infinity 16. Find the possible values of the common ratio of the progression.

8 Evaluate

   a) $\displaystyle\sum_{r=1}^{20} 5 \times 1.2^{r-1}$ b) $\displaystyle\sum_{r=1}^{\infty} 32 \times 0.6^{r-1}$ c) $\displaystyle\sum_{r=4}^{20} 2^r$

9 A sequence $v_1, v_2, v_3, \ldots$ is defined by

   $$v_1 = 32, \quad v_{n+1} = 1.25v_n.$$

   a) Calculate the values of v_2, v_3, v_4.
   b) Find the smallest value of n that satisfies $v_n > 1000$.
   c) Find $\displaystyle\sum_{r=1}^{50} v_r$.
   d) Find the smallest value of N that satisfies $\displaystyle\sum_{r=1}^{N} v_r > 10^6$.

10 A man is employed by a business on a contract which gives him a fixed 5% increase in salary each year.
   The increase takes effect from the first day of each year.
   He started work on 1st January 2002 on an annual salary of £24 000.
   If he stays working for this business
   a) what will his salary be in 2008?
   b) what will his **total** earnings be by the end of 2021?
   c) when will his **total** earnings first exceed £2 000 000?

11 Consider the geometric progression $\dfrac{37}{100} + \dfrac{37}{10\ 000} + \dfrac{37}{1\ 000\ 000} + \cdots$.
   a) Write down the sum S_n of the first n terms.
   b) What happens to the value of S_n as $n \to \infty$?
   c) Hence write the recurring decimal $0.373737373 \ldots$ as a fraction.
   d) Use a similar method to express the recurring decimal $0.241241241 \ldots$ as a fraction.

12 a) Find the sum of the first 20 terms of the progression

$$8 + 12 + 18 + 27 + \cdots$$

b) How many terms of this series must be added together to get a sum that exceeds 10^6?

13 On 1/1/04 Mr Wombat took out a mortgage with the Sydney Bay Building Society. The loan was for \$200 000 to be repaid over 25 years at a monthly interest rate of 0.5%. Mr Wombat repays the building society at a constant rate of \$$P$ per month on the last day of each month.

At the end of the first month of the loan Mr Wombat will therefore still owe the building society $200\,000 \times 1.005 - P$.

a) How much will he owe at the end of the second, third and fourth months?

b) Explain carefully why the amount owing at the end of the fifth month is

$$200\,000 \times 1.005^5 - P \times \frac{1.005^5 - 1}{0.005}$$

and write down a similar expression for the amount owing after n months.

c) Use the fact that the amount owing at the end of the 300th month is to be zero to determine the value of P and hence find the total of the repayments on this mortgage.

14 The Von Koch curve

Take a straight line.

Trisect the line and replace the central third by two sides of an equilateral triangle pointing outwards to give C_1.

Trisect each segment of C_1 and replace the central third by two sides of an equilateral triangle pointing outwards to give C_2.

Repeat to get C_3 and keep going.

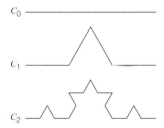

Von Koch's curve is the limit as $n \to \infty$ of C_n.

Let P_n denote the length of C_n and A_n denote the area inside C_n.

a) Show that $P_n = \left(\frac{4}{3}\right)^n P_0$.

b) Show that $A_n = \left(1 + \frac{4}{9} + \frac{16}{81} + \cdots + \left(\frac{4}{9}\right)^{n-1}\right) A_1$.

What can you deduce about the perimeter of, and the area inside, the Von Koch curve?

Having studied this chapter you should know

- the arithmetic progression

$$a, a + d, a + 2d, a + 3d, \ldots$$

has

$$n\text{th term} = a + (n - 1)d$$

and the sum of the first n terms, S_n, is given by

$$S_n = \frac{1}{2} n(a + L) = \frac{1}{2} n[2a + (n - 1)d]$$

where L denotes the last (nth) term;

- the sum of the first n natural numbers is $\frac{1}{2} n(n + 1)$
- the geometric progression

$$a, ar, ar^2, ar^3, \ldots$$

has

$$n\text{th term} = ar^{n-1}$$

and the sum of the first n terms, S_n, is given by

$$S_n = \frac{a(1 - r^n)}{1 - r} \quad \text{which can also be written as} \quad S_n = \frac{a(r^n - 1)}{r - 1}$$

Moreover, if $-1 < r < 1$ then the sum to infinity of the geometric progression exists and equals $\dfrac{a}{1 - r}$

REVISION EXERCISE

1 Find the sum of the following arithmetic progressions:
    a) $5 + 8 + 11 + \cdots + 95$ **b)** $72 + 68 + 64 + \cdots + (-16) + (-20)$

2 The ninth term of an arithmetic progression is 8 and the fourth term is 20. Find the first term and the common difference. How many of the terms are positive?

3 Show how the recurring decimal 0.235235235 ... may be written as a fraction.

4 Find the sum of the first 80 positive even integers.

5 A geometric progression $u_1, u_2, u_3, \ldots$ is defined by

$$u_1 = 600 \quad u_{n+1} = 0.8 \times u_n.$$

    a) Find the value of u_4.
    b) How many terms of the progression are greater than 1?
    c) What is the sum to infinity of the progression?

6 The first two terms of a geometric progression are 40 and 24.
Find
a) the eighth term, giving your answer correct to three decimal places;
b) the sum of the first ten terms, giving your answer correct to three decimal places;
c) the sum to infinity of the progression.

7 An arithmetic progression has third term 15 and the sum of the first twenty terms is 525.
Find
a) the first term and the common difference;
b) the number of terms that must be added together to obtain a sum greater than 15 000.

8 Prove that

a) $\displaystyle\sum_{r=1}^{10} (1 - a^2)a^r = a(1 + a)(1 - a^{10})$ **b)** $\displaystyle\sum_{r=1}^{n} \log(a^r) = \frac{1}{2} n(n + 1)\log a$

9 Find the smallest value of n such that the sum of the first n terms of the geometric progression

$$16 + 24 + 36 + \cdots$$

is greater than the sum of the first 1000 terms of the arithmetic progression

$$16 + 24 + 32 + \cdots$$

10 The first term of a sequence is 8 and the second term is 10.
i) Given that the terms of the sequence form an arithmetic progression, find the sum of the first 100 terms.
ii) Given instead that the terms of the sequence form a geometric progression and the sum of the first K terms is greater than 10^{15}, find the least possible value of K.

(OCR Jun 2001 P2)

11 A fitness programme includes a daily number of step-ups. The number of step-ups scheduled for day 1 is 20 and the number on each successive day is to be 3 more than on the previous day (i.e. 23 on day 2, 26 on day 3, etc.).
The programme also includes a daily time to be spent jogging. The time, T_N minutes, to be spent jogging on day N is given by the formula

$$T_N = 15 \times 1.05^{N-1}.$$

i) Find the total number of step-ups scheduled to be completed during the first 30 days of the fitness programme.
ii) Verify that the daily jogging programme first exceeds 60 minutes on day 30.
iii) Find the total time to be spent jogging during the first 30 days of the fitness programme.

(OCR Jan 2003 P2)

12 A sequence $v_1, v_2, v_3, \ldots$ is defined by

$$v_1 = 100, \quad v_{n+1} = v_n + 6.$$

a) Calculate the values of v_2, v_3, v_4.
b) Find the smallest value of n that satisfies $v_n > 1000$.
c) Find $\displaystyle\sum_{r=1}^{20} v_r$.
d) Find the smallest value of N that satisfies $\displaystyle\sum_{r=1}^{N} v_r > 10^6$.

8 Circular Measure

The purpose of this chapter is to enable you to

● use radians for angle measurement

● convert angle measurements between degrees and radians

● use formulae for arc length and sector area when the angle is expressed in radians

● use trigonometric functions for angles expressed in radians

● solve trigonometric equations with angles expressed in radians

Different Ways of Measuring Angles – Radians

In our everyday lives we are used to the fact that distances can be measured in miles or in kilometres, and you know that it is possible to convert between the two systems of units using one of the approximate relationships

$$5 \text{ miles} \approx 8 \text{ km} \qquad 1 \text{ mile} \approx 1.6 \text{ km} \qquad 0.625 \text{ miles} \approx 1 \text{ km}$$

So far, all the measurements of angles we've explored have used **degrees**. The division of a circle into 360 parts is thought to be due to the ancient Babylonian civilisation. Much more recently, the **radian** has been introduced as a unit for angular measurements. Although at first it may seem strange, the radian has many advantages over the degree in the study of mathematics. In this chapter we will see that formulae for arc length and sector area are much simpler if the angles are expressed in radians rather than degrees; in module C4 the real advantage of using radians will become apparent when the derivatives of trigonometric functions are investigated.

> **One radian** is the angle at the centre of a circle made by an arc whose length is the same as the radius of the circle.
>
> **Notation**
>
> An angle of 2.7 radians may be written as 2.7 rad.

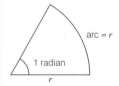

Conversion Between Degrees and Radians

If one radian is the same as $\alpha°$ then

$$r = \text{arc} = \frac{\alpha}{360} \times 2\pi r$$

$$\Rightarrow \quad 1 = \frac{\alpha}{360} \times 2\pi$$

$$\Rightarrow \quad 1 = \frac{\alpha}{180} \times \pi$$

$$\Rightarrow \quad \alpha = \frac{180}{\pi}.$$

> Recall that
>
> $$\text{arc length} = \frac{\text{angle of arc}}{\text{angle of whole circle}} \times \text{circumference of circle}$$
>
> $$= \frac{\alpha}{360} \times 2\pi r$$
>
> if the angles are measured in degrees.

So

$$1 \text{ radian} = \left(\frac{180}{\pi}\right)^{\circ}$$

This means that a radian is approximately 57.3°.

The basic conversion between radians and degrees can also be written as

$$1^{\circ} = \left(\frac{\pi}{180}\right) \text{radians}$$

The conversion between degrees and radians is most easily remembered as

$$180^{\circ} = \pi \text{ radians}$$

The previous two results are immediate consequences of this result.

Using these conversions, it can therefore be said that

$$2.7 \text{ rad} = 2.7 \times \frac{180}{\pi} = 154.698 \ldots^{\circ} = 154.7^{\circ} \quad \text{(to one decimal place)}$$

and

$$29^{\circ} = 29 \times \frac{\pi}{180} = \frac{29\pi}{180} = 0.50614 \ldots = 0.506 \text{ rad} \quad \text{(to three decimal places).}$$

Notice that angles such as 30°, 45°, 60°, 90°, etc. have neat **exact** conversions to radians, provided the answers are written as multiples of π.

For example

$$30^{\circ} = 30 \times \frac{\pi}{180} = \frac{30\pi}{180} = \frac{\pi}{6} \text{ rad} \qquad\qquad 120^{\circ} = 120 \times \frac{\pi}{180} = \frac{120\pi}{180} = \frac{2\pi}{3} \text{ rad}$$

$$270^{\circ} = 270 \times \frac{\pi}{180} = \frac{270\pi}{180} = \frac{3\pi}{2} \text{ rad} \qquad\qquad 360^{\circ} = 360 \times \frac{\pi}{180} = \frac{360\pi}{180} = 2\pi \text{ rad}$$

Formulae for Arc Length and Sector Area

If measuring in radians then

arc length = fraction of circle × $2\pi r$

$$= \frac{\theta}{2\pi} \times 2\pi r$$

$$= r\theta.$$

A whole circle makes an angle of 2π radians at its centre so the fraction of the circle made by an arc subtending θ radians at the centre is $\dfrac{\theta}{2\pi}$.

Similarly,

sector area = fraction of circle $\times \pi r^2$

$$= \frac{\theta}{2\pi} \times \pi r^2$$

$$= \frac{1}{2} r^2 \theta.$$

EXAMPLE 1

Find the perimeter and area of the sector of a circle of radius 5 cm that subtends an angle of 1.2 radians at the centre of the circle.

Arc $AB = r\theta = 5 \times 1.2 = 6$ cm.

Perimeter = 5 + 5 + arc AB = 16 cm.

Sector area $= \frac{1}{2} r^2 \theta = \frac{1}{2} \times 5^2 \times 1.2 = 15$ cm^2.

EXERCISE 1

1 Convert the following angles into degrees:

  a) 1.5 radians   **b)** 2.43 radians   **c)** 0.68 radians

  d) 0.07 radians   **e)** $\frac{\pi}{12}$ radians   **f)** $\frac{3\pi}{5}$ radians

2 i) Convert the following angles to radians, giving your answers correct to one decimal place:

    a) 32°   **b)** 106.3°   **c)** 222°

  ii) Convert the following angles to radians, giving your answer **exactly**, as a multiple of π.

    a) 60°   **b)** 150°   **c)** 300°

    d) 90°   **e)** 270°   **f)** 720°

3 Find the perimeter and area of this sector.

4 A sector of a circle has radius 6 cm and area 9 cm^2.
Find the angle, in radians, subtended at the centre of the circle by the arc.

5 The diagram shows a circular sector which has an area of 64 cm^2 and a perimeter of 32 cm.
Find the radius of the sector and the size of the angle θ.

6 A wheel of radius 2 metres is rotating at a rate of 80 revolutions per minute.
Find the speed of a point on the rim of the wheel.

Trigonometry with Angles Expressed in Radians

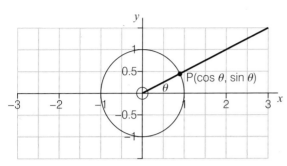

Recall from chapter 1 that $\cos \theta$ and $\sin \theta$ are defined as the x co-ordinate and the y co-ordinate, respectively, of the point P where a ray from the origin that makes an angle θ with the positive x-axis meets a circle of radius 1 unit and centre the origin.

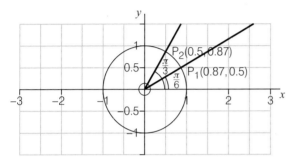

For example, the point P_1 is where the ray from the origin that makes an angle of $\dfrac{\pi}{6}$ radians (or 30°) with the x axis meets the circle:

$$\cos\left(\frac{\pi}{6}\right) = x_{P_1} \approx 0.87 \qquad \sin\left(\frac{\pi}{6}\right) = y_{P_1} \approx 0.5.$$

The point P_2 is where the ray from the origin that makes an angle of $\dfrac{\pi}{3}$ radians with the x axis meets the circle:

$$\cos\left(\frac{\pi}{3}\right) = x_{P_2} \approx 0.5 \qquad \sin\left(\frac{\pi}{3}\right) = y_{P_2} \approx 0.87.$$

Using these two results together with the symmetry of circle about the x and y axes, you can produce the following table of values:

θ (radians)	$\cos \theta$	$\sin \theta$		θ (radians)	$\cos \theta$	$\sin \theta$
0	1	0		$\dfrac{7\pi}{6}$	-0.87	-0.5
$\dfrac{\pi}{6}$	0.87	0.5		$\dfrac{4\pi}{3}$	-0.5	-0.87
$\dfrac{\pi}{3}$	0.5	0.87		$\dfrac{3\pi}{2}$	0	-1
$\dfrac{\pi}{2}$	0	1		$\dfrac{5\pi}{3}$	0.5	-0.87
$\dfrac{2\pi}{3}$	-0.5	0.87		$\dfrac{11\pi}{6}$	0.87	-0.5
$\dfrac{5\pi}{6}$	-0.87	0.5		2π	1	0
π	-1	0				

and plot the graphs of the two trigonometric functions:

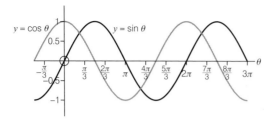

Notice that the each of the graphs repeats after 2π radians.

If the angles are measured in radians, then it is said that the cosine function and the sine function are periodic and have period 2π.

Notice that the graphs of $y = \sin\theta$ and $y = \cos\theta$ when the angles are measured in radians are exactly the same shape as the graphs of $y = \sin\theta$ and $y = \cos\theta$ when the angles are measured in degrees: the only thing that has changed is the markings on the horizontal scale.

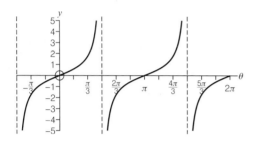

Similarly, the definition of $\tan\theta$ made in chapter 1 can be adapted for angles expressed in radians.

The graph of $y = \tan\theta$ when θ is measured in radians is the same as the graph of $y = \tan\theta$ when θ is measured in degrees except for the markings on the horizontal scale.

Notice that the tan function has period π radians and has vertical asymptotes at $-\dfrac{\pi}{2}, \dfrac{\pi}{2}, \dfrac{3\pi}{2}$, etc.

Exact Values of Trigonometric Ratios

In chapter 1 you saw that some angles have trigonometric ratios that may be expressed exactly. These are summarised in the table below for acute angles expressed both in degrees and in radians:

Angle (in degrees)	Angle (in radians)	cos(angle)	sin(angle)	tan(angle)
0	0	1	0	0
30	$\dfrac{\pi}{6}$	$\dfrac{\sqrt{3}}{2}$	$\dfrac{1}{2}$	$\dfrac{\sqrt{3}}{3}$
45	$\dfrac{\pi}{4}$	$\dfrac{\sqrt{2}}{2}$	$\dfrac{\sqrt{2}}{2}$	1
60	$\dfrac{\pi}{3}$	$\dfrac{1}{2}$	$\dfrac{\sqrt{3}}{2}$	$\sqrt{3}$
90	$\dfrac{\pi}{2}$	0	1	not defined

Using radians in calculations

Values of sin θ, cos θ and tan θ, where θ is in radians, can be obtained from your calculator once it has been switched into **radian mode**.

The trigonometric functions for angles expressed in radians can be used in exactly the same way as the trigonometric functions for angles expressed in degrees.

EXAMPLE 2

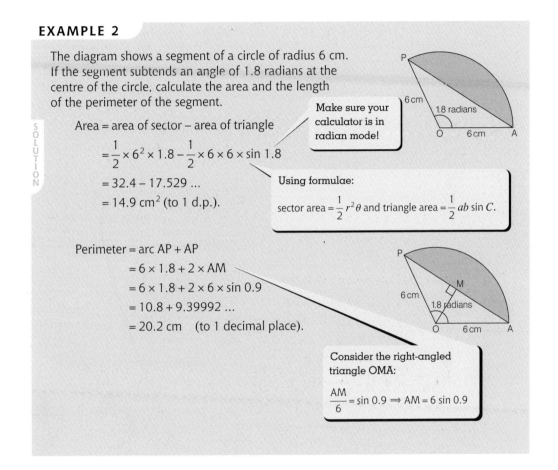

S
O
L
U
T
I
O
N

The diagram shows a segment of a circle of radius 6 cm. If the segment subtends an angle of 1.8 radians at the centre of the circle, calculate the area and the length of the perimeter of the segment.

Area = area of sector – area of triangle

Make sure your calculator is in radian mode!

$$= \frac{1}{2} \times 6^2 \times 1.8 - \frac{1}{2} \times 6 \times 6 \times \sin 1.8$$

$$= 32.4 - 17.529 \ldots$$

$$= 14.9 \text{ cm}^2 \text{ (to 1 d.p.)}.$$

Using formulae:

sector area $= \frac{1}{2} r^2 \theta$ and triangle area $= \frac{1}{2} ab \sin C$.

Perimeter = arc AP + AP

$$= 6 \times 1.8 + 2 \times AM$$

$$= 6 \times 1.8 + 2 \times 6 \times \sin 0.9$$

$$= 10.8 + 9.39992 \ldots$$

$$= 20.2 \text{ cm} \quad \text{(to 1 decimal place)}.$$

Consider the right-angled triangle OMA:

$$\frac{AM}{6} = \sin 0.9 \Rightarrow AM = 6 \sin 0.9$$

EXAMPLE 3

A triangle ABC has AB = 8 cm, BC = 14 cm and angle
ABC = 0.3 radians.
Calculate the length of AC.

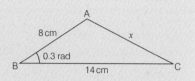

Using the cosine rule:

$$x^2 = 8^2 + 14^2 - 2 \times 8 \times 14 \cos 0.3$$
$$\Rightarrow \quad x^2 = 46.004 \ldots$$
$$\Rightarrow \quad x = 6.782 \ldots$$
$$\Rightarrow \quad x = 6.78 \text{ cm} \quad \text{(to 2 d.p.)}.$$

Make sure your calculator is in radian mode!

EXAMPLE 4

Solve the equation

$$\cos 2\theta = 0.8 \qquad 0 \leqslant \theta \leqslant 2\pi$$

giving your answers correct to two decimal places.

Remember that the calculator
needs to be in radian mode.

a) Start by putting $z = 2\theta$ to obtain the equation

$$\cos z = 0.8 \qquad 0 \leqslant z \leqslant 4\pi.$$

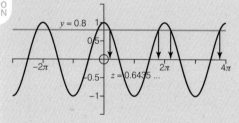

Using a calculator gives

$$z = \cos^{-1}(0.8) = 0.6435 \ldots$$

Using the symmetry of the cosine
graph, you also have

$$z = 2\pi - 0.6435 \ldots = 5.6396 \ldots$$

The fact that the cosine graph has period 2π means that the other two roots are

$$z = 2\pi + 0.6435 \ldots = 6.9266 \qquad \text{and} \qquad z = 2\pi + 5.6396 \ldots = 11.9228 \ldots$$

The roots of $\qquad \cos z = 0.8 \qquad 0 \leqslant z \leqslant 4\pi$

are $\qquad z = 0.6435 \ldots, 5.6396 \ldots, 6.9266 \ldots, 11.9288 \ldots$

But $z = 2\theta \quad \Rightarrow \quad \theta = \dfrac{z}{2} = \dfrac{0.6435 \ldots}{2} \text{ or } \dfrac{5.6396 \ldots}{2} \text{ or } \dfrac{6.9266 \ldots}{2} \text{ or } \dfrac{11.9228 \ldots}{2}$

$$\Rightarrow \quad \theta = 0.32 \text{ or } 2.82 \text{ or } 3.46 \text{ or } 5.96 \quad \text{(2 d.p.)}.$$

EXAMPLE 5

Find the **exact** roots of the equation

$$\sin 3\theta = \frac{1}{2} \qquad 0 \leqslant \theta \leqslant \pi$$

Start by putting $z = 3\theta$ to obtain the equation $\quad \sin z = \frac{1}{2} \qquad 0 \leqslant z \leqslant 3\pi.$

One solution is $z = \frac{\pi}{6}$.

> You know from chapter 1 that $\sin 30° = \frac{1}{2}$; the equivalent result
>
> for angles measured in radians is $\sin\left(\frac{\pi}{6}\right) = \frac{1}{2}$ so we know that
>
> $z = \frac{\pi}{6}$ is a solution of the equation.

Using the symmetry of the sine graph, you also have

$$z = \pi - \frac{\pi}{6} = \frac{5\pi}{6}.$$

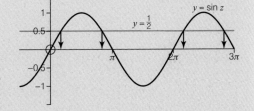

The fact that the sine graph has period 2π means that the other two roots are

$$z = 2\pi + \frac{\pi}{6} = \frac{13\pi}{6}$$

and

$$z = 2\pi + \frac{5\pi}{6} = \frac{17\pi}{6}.$$

The roots of $\quad \sin z = \frac{1}{2} \qquad 0 \leqslant z \leqslant 3\pi$

are

$$z = \frac{\pi}{6}, \frac{5\pi}{6}, \frac{13\pi}{6}, \frac{17\pi}{6}.$$

But $\quad z = 3\theta \quad \Rightarrow \quad \theta = \frac{z}{3} = \frac{\pi}{18}$ or $\frac{5\pi}{18}$ or $\frac{13\pi}{18}$ or $\frac{17\pi}{18}.$

EXAMPLE 6

Solve the equations:

a) $\sin\theta + 5\cos\theta = 0$
$0 \leqslant \theta \leqslant 2\pi$.

b) $6\sin^2\theta - \cos\theta - 4 = 0$
$0 \leqslant \theta \leqslant 2\pi$.

a)
$$\sin\theta + 5\cos\theta = 0$$
$$\Rightarrow \quad \sin\theta = -5\cos\theta$$
$$\Rightarrow \quad \tan\theta = -5$$

> Remember that $\tan\theta \equiv \dfrac{\sin\theta}{\cos\theta}$.

The calculator solution of this is

$$\theta = -1.3734 \ldots$$

which is outside the interval
$0 \leqslant \theta \leqslant 2\pi$, but remembering that
the tangent function has period π
gives the two possible values of θ:

$$\theta = -1.3734 + \pi = 1.7681 \ldots$$

or

$$\theta = -1.3734 + 2\pi = 4.9097 \ldots$$

so the roots are $\theta = 1.77$ or 4.91 (2 d.p.).

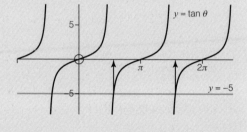

b)
$$6\sin^2\theta - \cos\theta - 4 = 0$$
$$\Rightarrow \quad 6(1 - \cos^2\theta) - \cos\theta - 4 = 0$$
$$\Rightarrow \quad 6 - 6\cos^2\theta - \cos\theta - 4 = 0$$
$$\Rightarrow \quad 2 - 6\cos^2\theta - \cos\theta = 0$$
$$\Rightarrow \quad 6\cos^2\theta + \cos\theta - 2 = 0$$
$$\Rightarrow \quad (3\cos\theta + 2)(2\cos\theta - 1) = 0$$
$$\Rightarrow \quad \cos\theta = -\frac{2}{3} \text{ or } \frac{1}{2}.$$

> Remember that $\sin^2\theta + \cos^2\theta \equiv 1$.
>
> This enables you to write $\sin^2\theta$ as
> $1 - \cos^2\theta$ and rewrite the equation as a
> quadratic equation in $\cos\theta$.

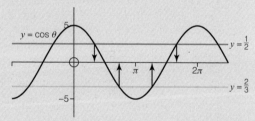

$$\cos\theta = -\frac{2}{3} \quad \Rightarrow \quad \theta = 2.3005 \ldots \text{ or } 2\pi - 2.3005 \ldots$$

$$\Rightarrow \quad \theta = 2.30 \text{ or } 3.98 \text{ (2 d.p.)}.$$

$$\cos\theta = \frac{1}{2} \quad \Rightarrow \quad \theta = \frac{\pi}{3} \text{ or } 2\pi - \frac{\pi}{3}$$

$$\Rightarrow \quad \theta = \frac{\pi}{3} \text{ or } \frac{5\pi}{3}.$$

> Remember that $\cos 60° = \dfrac{1}{2}$.
>
> The equivalent radian result is $\cos\left(\dfrac{\pi}{3}\right) = \dfrac{1}{2}$.

So the solution of the equation is

$$\theta = \frac{\pi}{3}, \frac{5\pi}{3}, 2.30 \text{ or } 3.98.$$

EXERCISE 2

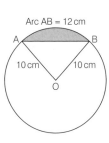

Arc AB = 12 cm

1 The arc AB of a circle of radius 10 cm and centre O has length 12 cm.
Find
a) the angle, in radians, subtended by the arc at O;
b) the length of the perimeter of the minor segment made by AB;
c) the area of the minor segment made by AB.

2 A triangle PQR has

  PQ = 10 cm   angle PQR = 0.4 rad   angle PRQ = 1.3 rad.

Calculate
i) the lengths of PR and RQ,
ii) the area of the triangle, giving your answers correct to two decimal places.

3 A triangle has sides of lengths 8 cm, 15 cm and 19 cm. Calculate the size, in radians, of the largest angle of the triangle. Find also the area of the triangle.

4 Sketch the graphs of
a) $y = \sin 2x$ **b)** $y = 2 + \cos x$ **c)** $y = -\tan x$

for values of x in the domain $0 \leqslant x \leqslant 2\pi$.

5 Find the exact solutions of the following equations, expressing your answers in radians:
a) $\tan x = 1$ $0 \leqslant x \leqslant 2\pi$
b) $\sin 2x = -0.5$ $0 \leqslant x \leqslant 2\pi$
c) $\cos 3x = 0$ $0 \leqslant x \leqslant 2\pi$

6 Find the **exact** solutions of the following equations:
a) $\cos 2\theta = -\frac{1}{2}$ $0 \leqslant \theta \leqslant 2\pi$
b) $\tan 3\theta = 1$ $-\pi \leqslant \theta \leqslant \pi$
c) $\tan^2 \theta = 3$ $0 \leqslant q \leqslant 2\pi$
d) $2 \cos \theta + 1 = 0$ $-\pi \leqslant \theta \leqslant \pi$
e) $2 \sin^2 \theta + \sin \theta - 1 = 0$ $-\pi \leqslant \theta \leqslant \pi$

7 Solve the following equations:
a) $2 \cos^2 \theta + \sin \theta - 1 = 0$ $-\pi \leqslant \theta \leqslant \pi$
b) $\tan \theta = -2 \sin \theta$ $0 \leqslant \theta \leqslant 2\pi$
c) $\cos^2 2\theta - \sin^2 2\theta = \dfrac{1}{2}$ $-\pi \leqslant \theta \leqslant \pi$

8 A chord PQ subtends an angle 2θ at the circumference of a circle with centre O and radius r.
a) Prove that triangle OPQ has area $r^2 \sin \theta \cos \theta$.
b) Find the area of the minor segment cut off by PQ.
c) If the area of the major segment formed by PQ is twice the area of the minor segment, show that

$$\theta - \sin \theta \cos \theta = \frac{\pi}{3}.$$

d) Using a trial and improvement method, or otherwise, find the value of θ correct to two decimal places.

Having studied this chapter you should know how to

- use radians to measure angles and be able to use the relationship

  π radians $= 180°$

  to convert angle measurements given in degrees to angle measurements given in radians and vice versa

- calculate arc length and sector area using the relationships

  $$\text{arc} = r\theta \qquad \text{sector area} = \frac{1}{2}r^2\theta$$

- sketch and use the graphs of the basic trigonometric functions when the angles are measured in radians

- solve trigonometric equations when the angles are measured in radians

REVISION EXERCISE

1   The diagram shows a sector of a circle with centre O and radius r.
    The arc AB subtends an angle of $105°$ at O.
    a)   Express $105°$ in radians, correct to three decimal places.
    b)   If the area of the sector is 25.74 cm^2, find the value of correct to one decimal place.

2   Solve the equations

   a) $\cos 2\theta = -\dfrac{1}{2}$ $0 \leqslant \theta \leqslant 2\pi.$

   b) $\sqrt{3}\sin\theta - \cos\theta = 0$ $0 \leqslant \theta \leqslant 2\pi.$

3   Sketch the graphs of

   a) $y = \cos 3\theta$ **b)** $y = 3 + \cos\theta$ **c)** $y = 2\cos\theta$

   for values of θ in the interval $0 \leqslant \theta \leqslant 2\pi.$

4   The diagram shows a sector of a circle of centre O and radius
    12 cm. The arc AB subtends an angle of 0.9 radians at O.
    P is the foot of the perpendicular from A to OB.

    Calculate the area of the shaded region.

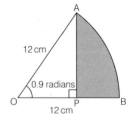

5   The diagram shows a golf green in the shape of two sectors.
    OA = 8 m, AB = 5 m.
    Calculate the area of the green.

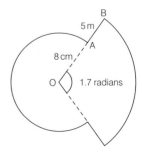

6 a) Find the exact roots of the equation $4\cos^2\theta = 1$ in the interval $-\pi \leqslant \theta \leqslant \pi$.

b) Show that $4\sin^4\theta + 7\cos^2\theta - 4 \equiv \cos^2\theta(4\cos^2\theta - 1)$ and hence solve the equation $4\sin^4\theta + 7\cos^2\theta - 4 = 0$, $-\pi \leqslant \theta \leqslant \pi$.

7 Triangle ABC is such that AB = 10 cm, BC = 8 cm and CA = 7 cm. A circular arc with centre A and radius 5 cm is drawn from the point D on AB to the point E on AC. (See diagram.)

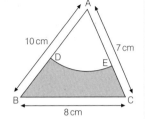

i) Show that, correct to 2 decimal places, angle BAC = 0.92 radians.

ii) Find the perimeter and area of the shaded region BDEC.

8 Show that the equation $15\sin^2\theta + \cos\theta = 13$ can be written as

$$15\cos^2\theta - \cos\theta - 2 = 0.$$

Hence find all roots of the equation $15\sin^2\theta + \cos\theta = 13$ in the interval $0 \leqslant \theta \leqslant 2\pi$.

9 The diagram shows the graph of $y = \tan x$ for $-\dfrac{\pi}{2} < x < 2\pi$ and the graph of a second trigonometric function.

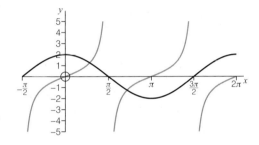

a) Write down the equation of the second graph.

b) Find, correct to three decimal places, the co-ordinates of the points of intersection of the two graphs.

10 A children's roundabout is circular and rotates at a constant rate of 0.4 radians/second.

a) How long does the roundabout take to turn through one complete revolution?

The roundabout has radius 1.8 m. Stephen stands on the edge of the roundabout.

b) At what speed is Stephen moving?

Stephen's mother is standing by the side of the roundabout. At $t = 0$ Stephen and his mother are next to each other. A short time later, at time t seconds, Stephen's position is such that $\angle MOS = 0.4t$ radians.

Initially | Time t

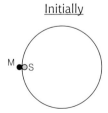 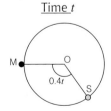

c) Prove that $MS = 3.6\sin 0.2t$.

d) Sketch a graph to show how MS varies with t for $0 \leqslant t \leqslant 5\pi$.

e) Find the values of t during the first revolution of the roundabout for which the distance between Stephen and his mother is 3 m.

9 Pascal's Triangle and the Binomial Expansion

The purpose of this chapter is to enable you to

● use Pascal's triangle or combinations to write down the expansion of $(a + b)^n$, for positive integer values of n.

Pascal's Triangle

Pascal's triangle is the number pattern whose first few lines are

$$
\begin{array}{ccccccccccccc}
 & & & & & 1 & & 1 & & & & & \\
 & & & & 1 & & 2 & & 1 & & & & \\
 & & & 1 & & 3 & & 3 & & 1 & & & \\
 & & 1 & & 4 & & 6 & & 4 & & 1 & & \\
 & 1 & & 5 & & 10 & & 10 & & 5 & & 1 & \\
1 & & 6 & & 15 & & 20 & & 15 & & 6 & & 1 \\
\end{array}
$$

Notice how each number in a line is the sum of the two numbers immediately above it in the previous line.

> In this exercise some of the properties of Pascal's triangle are investigated. Before you proceed to the next section make sure that you have answered at least questions 1, 2, 5 and 6.

EXERCISE 1

1 Write down the next three lines of the pattern.

2 Add the numbers in each line of the pattern.
What do you notice?
What do you think the sum of the numbers in the fifteenth row will be?

3 1, 3, 6, 10 are the first four TRIANGULAR numbers.

Write down the next four triangular numbers.
Find a formula for the nth triangular number.
How are the triangular numbers related to the numbers in Pascal's triangle?

```
    O
   O O
  O O O
 O O O O
```

4 The map shows part of the road network in an American city.

To reach A from O in the minimum distance there is a choice of three possible routes: EEN or ENE or NEE.

To reach B from O in the minimum distance there are 15 possible routes. What are they?

Copy the diagram and fill in at each junction the number of minimum length routes from O to that junction.

What do you notice?

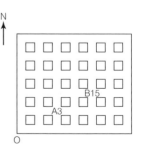

5 a) Expand $(1 + x)^2$.

b) Expand $(1 + x)^3$.

c) Prove that $(1 + x)^4 = 1 + 4x + 6x^2 + 4x^3 + x^4$.

d) What do you think the expansions of $(1 + x)^5$, $(1 + x)^6$ and $(1 + x)^7$ will be? If possible, use a computer algebra system to check your answers.

6 a) Expand $(x + y)^2$.

b) Expand $(x + y)^3$.

c) Prove that $(x + y)^4 = x^4 + 4x^3 y + 6x^2 y^2 + 4xy^3 + y^4$.

d) What do you think the expansions of $(x + y)^5$, $(p + q)^6$ and $(a + b)^7$ will be? If possible, use a computer algebra system to check your answers.

Using Pascal's Triangle

In question 6 of exercise 1 you saw that

$$(x + y)^2 = \mathbf{1}x^2 + \mathbf{2}xy + \mathbf{1}y^2$$
$$(x + y)^3 = \mathbf{1}x^3 + \mathbf{3}x^2 y + \mathbf{3}xy^2 + \mathbf{1}y^3$$
$$(x + y)^4 = \mathbf{1}x^4 + \mathbf{4}x^3 y + \mathbf{6}x^2 y^2 + \mathbf{4}xy^3 + \mathbf{1}y^2.$$

You can see that the numbers in the kth row of Pascal's triangle give the coefficients of the terms in the expansion of $(x + y)^k$.

You can use Pascal's triangle to quickly write down the expansion of expressions such as $(p + q)^6$.

- The sixth row of Pascal's triangle is 1, 6, 15, 20, 15, 6, 1.
- The first term will be p^6; in each subsequent terms the power of p decreases by 1 and the power of q increases by 1 so the subsequent terms are $p^5 q$, $p^4 q^2$, $p^3 q^3$, $p^2 q^4$, pq^5, q^6.

You can write

$$(p + q)^6 = 1p^6 + 6p^5 q + 15p^4 q^2 + 20p^3 q^3 + 15p^2 q^4 + 6pq^5 + 1q^6$$
$$= p^6 + 6p^5 q + 15p^4 q^2 + 20p^3 q^3 + 15p^2 q^4 + 6pq^5 + q^6$$

With care, you can also use Pascal's triangle to quickly write down the expansions of more complicated expressions.

EXAMPLE 1

Find the expansion of $(2x + 3y)^4$.

The fourth row of Pascal's triangle is 1, 4, 6, 4, 1 so

$$(p + q)^4 = p^4 + 4p^3 q + 6p^2 q^2 + 4pq^3 + q^4.$$

Replacing p by $2x$ and q by $3y$ gives

$$(2x + 3y)^4 = (2x)^4 + 4(2x)^3(3y) + 6(2x)^2(3y)^2 + 4(2x)(3y)^3 + (3y)^4$$
$$\Rightarrow \quad (2x + 3y)^4 = 16x^4 + 96x^3 y + 216x^2 y^2 + 216xy^3 + 81y^4.$$

Great care is needed here. For example, remember that

$$6(2x)^2(3y)^2 = 6 \times 2^2 x^2 \times 3^2 y^2 = 6 \times 4x^2 \times 9y^2 = 216x^2 y^2.$$

EXAMPLE 2

Find the expansion of $(2 - 3x)^5$ and hence find the exact value of 1.97^5.

The fifth row of Pascal's triangle is 1, 5, 10, 10, 5, 1 so

$$(p + q)^5 = p^5 + 5p^4q + 10p^3q^2 + 10p^2q^3 + 5pq^4 + q^5.$$

Replacing p by 2 and q by $-3x$ gives

$$(2 - 3x)^5 = (2)^5 + 5(2)^4(-3x) + 10(2)^3(-3x)^2 + 10(2)^2(-3x)^3 + 5(2)(-3x)^4 + (-3x)^5$$
$$\Rightarrow \quad (2 - 3x)^5 = 32 - 240x + 720x^2 - 1080x^3 + 810x^4 - 243x^5.$$

> Again, care is needed at this stage. For example, remember that
> $$10(2)^2(-3x)^3 = 10 \times 2^2 \times (-3)^3x^3 = 10 \times 4 \times -27x^3 = -1080x^3.$$

Replacing x by 0.01 now gives

$$1.97^5 = 32 - 2.4 + 0.072 - 0.001080 + 0.00000810 - 0.0000000243$$
$$\Rightarrow \quad 1.97^5 = 32 + 0.072 + 0.00000810 - 2.4 - 0.001080 - 0.0000000243$$
$$\Rightarrow \quad 1.97^5 = 32.0720081000 - 2.4010800243 = 29.6709280757.$$

> Why is this different from the answer your calculator gives?

EXAMPLE 3

If x is so small that x^3 and higher powers can be ignored, find the expansion of $(1 - 2x)(3 - 4x)^5$.

Using Pascal's triangle

$$(p + q)^5 = p^5 + 5p^4q + 10p^3q^2 + 10p^2q^3 + 5pq^4 + q^5$$

Replacing p by 3 and q by $-4x$ gives

$$(3 - 4x)^5 = 3^5 + 5 \times 3^4(-4x) + 10 \times 3^3(-4x)^2 + 10 \times 3^2(-4x)^3 + 5 \times 3(-4x)^4 + (-4x)^5$$
$$= 243 - 1620x + 4320x^2 + ...$$

> The remaining terms are omitted since x is so small x^3 and higher powers can be ignored.

$$(1 - 2x)(3 - 4x)^5 = (1 - 2x)(243 - 1620x + 4320x^2 + ...)$$
$$= 243 - 1620x + 4320x^2 + ...$$
$$- 486x + 3240x^2 + ...$$
$$= 243 - 2106x + 7560x^2 + ...$$

> The remaining terms are omitted since x is so small x^3 and higher powers can be ignored.

EXAMPLE 4

Find the expansion of $\left(u^2 + \dfrac{2}{u}\right)^5$.

Using Pascal's triangle

$$(p+q)^5 = p^5 + 5p^4q + 10p^3q^2 + 10p^2q^3 + 5pq^4 + q^5.$$

Replacing p by u^2 and q by $\dfrac{2}{u}$ gives

$$\left(u^2 + \frac{2}{u}\right)^5 = (u^2)^5 + 5(u^2)^4\left(\frac{2}{u}\right) + 10(u^2)^3\left(\frac{2}{u}\right)^2 + 10(u^2)^2\left(\frac{2}{u}\right)^3 + 5(u^2)\left(\frac{2}{u}\right)^4 + \left(\frac{2}{u}\right)^5$$

$$= u^{10} + 5 \times u^8 \times \frac{2}{u} + 10 \times u^6 \times \frac{4}{u^2} + 10 \times u^4 \times \frac{8}{u^3} + 5 \times u^2 \times \frac{16}{u^4} + \frac{32}{u^5}$$

$$= u^{10} + 10u^7 + 40u^4 + 80u + \frac{80}{u^2} + \frac{32}{u^5}.$$

EXERCISE 2

1 Expand   **a)** $(a+b)^6$ **b)** $(3+2x)^5$ **c)** $(5-4x)^3$

2 Find the expansions of

   a) $\left(x + \dfrac{4}{x}\right)^3$ **b)** $\left(y^2 - \dfrac{2}{y}\right)^4$ **c)** $\left(z + \dfrac{1}{z}\right)^6$

3 Write down the expansion of $(3+x)^7$. Hence find the exact value of 3.1^7.

4 Find the expansions of   **a)** $(4+x)(2+3x)^3$ **b)** $(1-3x)(3-x)^4$

5 If x is so small that x^4 and higher powers can be ignored, show that

$$(3+2x)(1-x)^9 \approx 3 - 25x + 90x^2 - 180x^3.$$

6 If x is so small that x^3 and higher powers can be ignored
   i)   find the expansion of $(2-5x)^6$,
   ii)   given that the expansion of $(\alpha + \beta x)(2-5x)^6$ is

$$192 - 2624x + \gamma x^2 + \ldots$$

   find the values of the constants α, β and γ.

7 Find the expansion of $\left(z^3 + \dfrac{5}{z}\right)^4 - \left(z^3 - \dfrac{5}{z}\right)^4$.

8 Prove that $(\sqrt{5} - \sqrt{2})^5 = 145\sqrt{5} - 229\sqrt{2}$.

9 If x is so small that x^3 and higher powers are negligible, show that

$$(5 - 2x)(1 + 3x)^5 = 4 + 73 + px^2$$

where p is a constant whose value should be stated.
Hence, without using a calculator, obtain an estimate of the value of 4.98×1.03^5.

10 a) Write out the expansion of $(p + qx)^4$.

   b) If x is so small that x^3 and higher powers are negligible then the expansion of $(\alpha + \beta x)(2 + \gamma x)^4$ is $64 + 592x + 1920x^2$. Find the values of the positive constants α, β and γ.

Formalising Pascal's Triangle

If you wanted to expand expressions such as $(3 + 2x)^{15}$ you would need to work out the fifteenth row of Pascal's triangle – a very time consuming exercise if you have to write out all 15 rows. What is needed is a quick way of working out the numbers in the fifteenth (or any other) row without writing out all the earlier rows.

In this section you will look closely at Pascal's triangle and find ways of predicting what numbers in the nth row will be.

- First, note that the first number in each row of Pascal's triangle is 1.
- Now look at the second number in each row of the triangle.

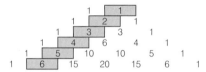

You would expect the second number in the nth row of Pascal's triangle to be n.

- Now look at the third number in each row of the triangle.

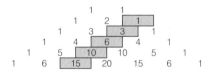

Consider the ratio of these numbers to the numbers which appear immediately to the left of them in the triangle.

You can see that the third number in the nth row appears to be $\dfrac{n-1}{2}$ times the second number in the row. This together with the previous result leads to the conjecture that

$$\text{3rd number in } n\text{th row} = \text{2nd number in } n\text{th row} \times \frac{n-1}{2}$$

$$= n \times \frac{n-1}{2}.$$

● Now look at the fourth number in each row of the triangle.

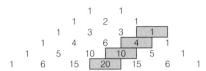

Consider the ratio of these numbers to the numbers which appear immediately to the left of them in the triangle.

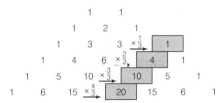

You can see that the fourth number in the nth row appears to be $\dfrac{n-2}{3}$ times the second number in the row. This together with the previous result leads to the conjecture that

$$\text{4th number in } n\text{th row} = \text{3rd number in } n\text{th row} \times \dfrac{n-2}{3}$$

$$= n \times \dfrac{n-1}{2} \times \dfrac{n-2}{3}.$$

● Using a similar method, you can obtain predictions for the fifth and sixth numbers in the nth row of Pascal's triangle:

$$\text{5th number in } n\text{th row} = \text{4th number in } n\text{th row} \times \dfrac{n-3}{4}$$

$$= n \times \dfrac{n-1}{2} \times \dfrac{n-2}{3} \times \dfrac{n-3}{4}$$

and

$$\text{6th number in } n\text{th row} = \text{5th number in } n\text{th row} \times \dfrac{n-4}{5}$$

$$= n \times \dfrac{n-1}{2} \times \dfrac{n-2}{3} \times \dfrac{n-3}{4} \times \dfrac{n-4}{5}$$

> What do you think the rule for the seventh number in the nth row of Pascal's triangle will be?

The Factorial Function

You have now got some rather cumbersome rules for the numbers in Pascal's triangle – you will need a shorthand which will make them rather easier to handle.

Definition: If n is a positive integer then n factorial, written $n!$, is the product of all the whole numbers from 1 to n.

$$n! = 1 \times 2 \times 3 \times 4 \times \cdots \times (n-1) \times n$$

where $0!$ is defined to be 1.

Thus
$$0! = 1$$
$$1! = 1$$
$$2! = 1 \times 2 = 2$$
$$3! = 1 \times 2 \times 3 = 6$$
$$4! = 1 \times 2 \times 3 \times 4 = 24$$
$$5! = 1 \times 2 \times 3 \times 4 \times 5 = 120$$
$$6! = 1 \times 2 \times 3 \times 4 \times 5 \times 6 = 720, \text{ etc.}$$

Many numbers can be written neatly in terms of factorials. Consider for example the numbers

$$5 \times 6 \times 7 \times 8 \times 9 = \frac{1 \times 2 \times 3 \times 4 \times 5 \times 6 \times 7 \times 8 \times 9}{1 \times 2 \times 3 \times 4} = \frac{9!}{4!}$$

$$2 \times 4 \times 6 \times 8 \times 10 = (2 \times 1) \times (2 \times 2) \times (2 \times 3) \times (2 \times 4) \times (2 \times 5)$$
$$= 2^5 \times (1 \times 2 \times 3 \times 4 \times 5) = 2^5 \times 5!$$

You have already seen that the first six numbers in the nth row of Pascal's triangle are

$$1, n, \frac{n(n-1)}{2}, \frac{n(n-1)(n-2)}{6}, \frac{n(n-1)(n-2)(n-3)}{24}, \frac{n(n-1)(n-2)(n-3)(n-4)}{120}.$$

Consider the sixth of these numbers:

you know that $5! = 120$

and

$$n(n-1)(n-2)(n-3)(n-4) = \frac{n(n-1)(n-2)(n-3)(n-4) \times (n-5)(n-6) \times \cdots \times 2 \times 1}{(n-5)(n-6) \dots \times 2 \times 1} = \frac{n!}{(n-5)!}$$

so that

$$\text{sixth number in } n\text{th row of Pascal's triangle} = \frac{n(n-1)(n-2)(n-3)(n-4)}{120}$$

$$= \frac{\frac{n!}{(n-5)!}}{5!}$$

$$= \frac{n!}{(n-5)!\,5!}$$

In a similar way, you can show that

$$\text{fifth number in } n\text{th row of Pascal's triangle} = \frac{n!}{(n-4)!\,4!}$$

$$\text{fourth number in } n\text{th row of Pascal's triangle} = \frac{n!}{(n-3)!\,3!}$$

$$\text{third number in } n\text{th row of Pascal's triangle} = \frac{n!}{(n-2)!\,2!}$$

If you have worked on the S1 module you may have already met $_nC_r$ as the number of different ways of choosing r items from n when the order of choice is unimportant.

Your calculator should enable you to calculate these numbers directly.

Notation: The number $_nC_r$ or $\binom{n}{r}$ is shorthand for $\frac{n!}{(n-r)!\,r!}$.

Notice that

$$_nC_0 = \frac{n!}{(n-0)!\,0!} = \frac{n!}{n! \times 1} = 1 = \text{first number in } n\text{th row of Pascal's triangle}$$

and

$$_nC_1 = \frac{n!}{(n-1)!\,1!} = \frac{n!}{(n-1)! \times 1} = \frac{n!}{(n-1)!} = n = \text{second number in } n\text{th row of Pascal's triangle.}$$

Putting all these results together, you know that

- the numbers, or coefficients, in the expansion of $(x+y)^n$ are the numbers in the nth row of Pascal's triangle;
- the numbers in the nth row of Pascal's triangle are $_nC_0, {}_nC_1, {}_nC_2, {}_nC_3, \ldots, {}_nC_r, \ldots, {}_nC_n$

so you can conclude that

$$(x+y)^n = {}_nC_0 x^n + {}_nC_1 x^{n-1} y + {}_nC_2 x^{n-2} y^2 + {}_nC_3 x^{n-3} y^3 + \cdots + {}_nC_r x^{n-r} y^r + \cdots + {}_nC_n y^n$$

and, remembering that $_nC_0 = {}_nC_n = 1$, produces the final result:

$$(x+y)^n = x^n + {}_nC_1 x^{n-1} y + {}_nC_2 x^{n-2} y^2 + {}_nC_3 x^{n-3} y^3 + \cdots + {}_nC_r x^{n-r} y^r + \cdots + y^n.$$

Using the $\binom{n}{r}$ notation this result can be written as

$$(x+y)^n = x^n + \binom{n}{1} x^{n-1} y + \binom{n}{2} x^{n-2} y^2 + \binom{n}{3} x^{n-3} y^3 + \cdots + \binom{n}{r} x^{n-r} y^r + \cdots + y^n.$$

EXAMPLE 5

Find the first four terms in the expansion of $\left(2 - \frac{1}{2}x\right)^{15}$.

$$(a+b)^{15} = a^{15} + {}_{15}C_1 a^{14} b + {}_{15}C_2 a^{13} b^2 + {}_{15}C_3 a^{12} b^3 + \cdots$$
$$\Rightarrow \quad (a+b)^{15} = a^{15} + 15a^{14} b + 105a^{13} b^2 + 455a^{12} b^3 + \cdots.$$

Putting $a = 2$ and $b = -\frac{1}{2}x$ gives:

$$\left(2 - \frac{1}{2}x\right)^{15} = 2^{15} + 15 \times 2^{14} \times \left(-\frac{1}{2}x\right) + 105 \times 2^{13} \times \left(-\frac{1}{2}x\right)^2 + 455 \times 2^{12} \times \left(-\frac{1}{2}x\right)^3 + \cdots.$$

$$\Rightarrow \quad \left(2 - \frac{1}{2}x\right)^{15} = 32\,768 - 122\,880x + 215\,040x^2 - 232\,960x^3 + \cdots.$$

Great care is again needed at this stage.
For example, remember that

$$105 \times 2^{13} \times \left(-\frac{1}{2}x\right)^2 = 105 \times 2^{13} \times \left(-\frac{1}{2}\right)^2 x^2$$

$$= 105 \times 2^{13} \times \frac{1}{4}x^2 = 215\,040x^2.$$

EXAMPLE 6

Find the x^8 term in the expansion of $\left(x^4 + \dfrac{1}{x^2}\right)^{20}$.

The general term in the expansion of $(a+b)^{20}$ is of the form $_{20}C_r a^{20-r} b^r$.

The general term in the expansion of $\left(x^4 + \dfrac{1}{x^2}\right)^{20}$ is therefore of the form

$$_{20}C_r (x^4)^{20-r} \left(\frac{1}{x^2}\right)^r.$$

Now

$$_{20}C_r (x^4)^{20-r} \left(\frac{1}{x^2}\right)^r = {}_{20}C_r x^{80-4r} \frac{1}{x^{2r}} = {}_{20}C_r x^{80-6r}.$$

This will be an x^8 term if $80 - 6r = 8 \implies r = 12.$

The x^8 term is therefore $_{20}C_{12} x^8 = 125\ 970 x^8.$

EXERCISE 3

1 Find the first four terms of the expansions of

  a) $(1+y)^{16}$ **b)** $(2-x)^{10}$ **c)** $(1+3x)^{14}$

2 Find the first four terms of the expansion of $(1+7x)^{12}$.
Hence, without further use of your calculator, obtain an estimate of the value of 1.07^{12}.
State, with reasons, whether your answer is an under- or over-estimate of the correct value.

3 Find the   **a)** x^5 term in the expansion of $\left(x + \dfrac{2}{x}\right)^9$

           b) x^{-6} term in the expansion of $\left(x^2 - \dfrac{3}{x}\right)^{15}$.

4 In the expansion of $(1 - 2ax)^{15}$ in powers of x, the coefficient of x^2 is $\dfrac{28}{15}$.
Find the possible values of a.
If a is positive, find the first four terms of the expansion.

5 If x is so small that x^5 and higher powers can be ignored, prove that

$$(1+4x)^{12} - (1-4x)^{12} = 96x + 28\ 160x^3.$$

Hence, without using a calculator, estimate the value of $1.004^{12} - 0.996^{12}$.

6 If y is so small that y^3 and higher powers can be ignored, obtain the expansion of

$$(5+4y)\left(1 - \frac{1}{2}y\right)^{16}.$$

7 If u is so small that u^5 and higher powers can be ignored

a) prove that $\left(2 + \dfrac{1}{4}u\right)^{10} = 1024 + 1280u + 720u^2 + 240u^3 + 52.5u^4$;

b) deduce the expansion of $\left(2 + \dfrac{1}{4}u^2\right)^{10}$;

c) obtain the expansion of $\left(2 + \dfrac{1}{4}u^2\right)^{10} + \left(2 - \dfrac{1}{4}u^2\right)^{10}$.

8 Find the value of the constant a if, in the expansion of $(a + x)^9$, the coefficient of x^3 is twice the coefficient of x^4.

Having studied this chapter you should know

● the first few lines of Pascal's triangle

$$
\begin{array}{ccccccccccccc}
 & & & & & 1 & & 1 & & & & & \\
 & & & & 1 & & 2 & & 1 & & & & \\
 & & & 1 & & 3 & & 3 & & 1 & & & \\
 & & 1 & & 4 & & 6 & & 4 & & 1 & & \\
 & 1 & & 5 & & 10 & & 10 & & 5 & & 1 & \\
1 & & 6 & & 15 & & 20 & & 15 & & 6 & & 1 \\
\end{array}
$$

and realise that the nth row gives the coefficients of the terms in the expansion of $(p + q)^n$

● that $n! = 1 \times 2 \times 3 \times 4 \times \cdots \times (n - 1) \times n$ and $0! = 1$

● that $\dbinom{n}{r} = {}_nC_r = \dfrac{n!}{(n - r)!\,r!}$

● the numbers in the nth row of Pascal's triangle are ${}_nC_0, {}_nC_1, {}_nC_2, {}_nC_3, \ldots, {}_nC_r, \ldots, {}_nC_n$

● $(x + y)^n = x^n + {}_nC_1 x^{n-1}y + {}_nC_2 x^{n-2}y^2 + {}_nC_3 x^{n-3}y^3 + \cdots + {}_nC_r x^{n-r}y^r + \cdots + y^n$

or

$(x + y)^n = x^n + \dbinom{n}{1}x^{n-1}y + \dbinom{n}{2}x^{n-2}y^2 + \dbinom{n}{3}x^{n-3}y^3 + \cdots + \dbinom{n}{r}x^{n-r}y^r + \cdots + y^n$

REVISION EXERCISE

1 a) Write down the expansion of $(a + b)^4$.

b) Prove that $(3 + \sqrt{2})^4 = 193 + 132\sqrt{2}$.

2 a) Find the expansion of $(1 + 2y)^5$.

b) Deduce the expansion of $(1 - 2y)^5$.

c) Hence solve the equation $(1 + 2y)^5 + (1 - 2y)^5 = 8402$.

3 a) Expand $(3 - 2x)^5$ in ascending powers of x up to and including the x^2 term, simplifying the coefficients.

b) In the expansion of $(5 + ax)(3 - 2x)^5$ the coefficient of x^2 is 2160. Find the value of the constant a.

4 **a)** Show that $(4 + x)^3 - x^3 - 75 \equiv 12x^2 + 48x - 11$.

    b) Hence find the quotient and remainder when $(4 + x)^3 - x^3 - 75$ is divided by $(x + 2)$.

5 **a)** Given that the first three terms in the expansion of $(2 - x^2)^5$ are $p + qx^2 + rx^4$, find the values of the constants p, q and r.

    b) Hence find the coefficient of x^4 in the expansion of $(1 + 3x^2 - x^4)(2 - x^2)^5$.

6 **a)** Find the expansions of $(3 + 2x^5)^3$ and $(3 + 2x^5)^4$.

    b) Hence prove that if $y = (3 + 2x^5)^4$ then $\dfrac{dy}{dx} = 40x^4(3 + 2x^5)^3$.

7 The value of the definite integral $\displaystyle\int_0^{0.1} (1 + 3x^2)^{12} \, dx$ is denoted by J.

    i) Use the trapezium rule with two intervals, each of width 0.05, to find an approximation to J, giving your answer correct to 4 decimal places.

    A second approximation to J is to be found based on a binomial expansion.

    ii) The first three terms in the expansion of $(1 + 3x^2)^{12}$ are $1 + ax^2 + bx^4$. Find the values of the constants a and b.

    iii) Hence, by evaluating $\displaystyle\int_0^{0.1} (1 + ax^2 + bx^4) \, dx$, find an approximation to J, giving your answer correct to 4 decimal places.

(OCR Jan 2003 P2)

8 Prove that

$$(2 + 3x)^5 - (2 - 3x)^5 \equiv 480x + 2160x^3 + 486x^5.$$

Without using a calculator
    i) deduce the exact value of $2.3^5 - 1.7^5$,

    ii) prove that $(2 + 3\sqrt{2})^5 - (2 - 3\sqrt{2})^5 = p\sqrt{2}$ where p is a positive integer whose value should be stated.

9 The first three terms in the expansion of $(2 + px)^n$, where n is a positive integer, are

$$128 - 2240x + qx^2 + \cdots .$$

Obtain the values of n, p and q.

10 In the expansion of $(5 + 2x)^n$ the coefficients of x and x^2 are equal. Determine the value of n.

Revise chapters 1, 2 and 8 before attempting this exercise.

1 The diagram represents four points A, B, C and D
on horizontal ground. ACD is a straight line, angle
A = 25°, AC = 300 m and BC = BD = 500 m.

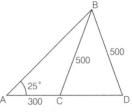

   a) Calculate angle ABC.

   b) Calculate the distance CD.

(OCR Jun 1995 P1)

2 The diagram shows a sector of a circle with centre
O and radius 6 cm. Angle POQ = 0.6 radians.

Calculate the length of arc PQ and the area of the
sector OPQ.

(OCR Nov 1995 P1)

3 On a single diagram, for $0 \leqslant x < \dfrac{3}{2}\pi$, sketch the graphs of

   i) $y = \tan x$

   ii) $y = \dfrac{1}{2}\pi - x$

including on your sketch the co-ordinates, in terms of π, of the points of intersection of
each of these graphs with the axes.

How many roots does the equation $\tan x = \dfrac{1}{2}\pi - x$ have in the interval $0 \leqslant x < \dfrac{3}{2}\pi$?

(OCR Nov 1997 P1, adapted)

4 Giving your answers exactly, find all the roots of the equations

   a) $\cos x = \dfrac{\sqrt{3}}{2}$ $\qquad 0 \leqslant x \leqslant 4\pi$

   b) $\sin 2x = -\dfrac{1}{2}$ $\qquad -180° \leqslant x \leqslant 180°$

   c) $\tan 3x = -\sqrt{3}$ $\qquad 0 \leqslant x \leqslant \pi$

5 The diagram shows a triangle ABC in which
angle C = 30°, BC = x cm and AC = $(x + 2)$ cm.
Given that the area of the triangle ABC is
12 cm², calculate the value of x. Find also the
length of AB, giving your answer correct to
two decimal places.

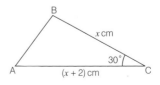

(OCR Jun 1996 P1, adapted)

6 The diagram shows a sector of a circle with centre O and radius r cm. The arc ACB subtends an angle of 259° at O.

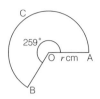

a) Express 259° in radians.

b) Given that the area of the sector is 114 cm², find the value of r.

(OCR Mar 1998 P1)

7 The diagram shows the graph of $y = \cos x$, where the angle x is measured in degrees.

Write down the co-ordinates of the points S and T.

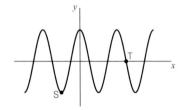

8 Without using a calculator, determine the exact value of $(\cos 30° + 4 \sin 30°)^2$.

9 i) Show that the equation $3 \sin 2\theta - \cos 2\theta = 0$ may be written as $\tan 2\theta = \dfrac{1}{3}$.

ii) Hence solve the equation $3 \sin 2\theta - \cos 2\theta = 0$, giving all values of θ such that $0° \leqslant \theta \leqslant 360°$. Give your answers to the nearest 0.1°.

(OCR Jan 2001 P1)

10 The diagram shows a triangle ABC, and a semicircle with BC as diameter. Angle A is 60° and the lengths of AC, AB and BC are p, $p + 5$ and q centimetres, respectively.

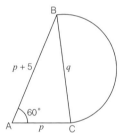

a) Express q^2 in terms of p and hence show that the area of the semicircle is

$$\frac{1}{8} \pi(p^2 + 5p + 25) \text{ cm}^2.$$

b) The area of the triangle ABC is equal to the area of the semicircle. Obtain and simplify an equation for p, giving the coefficients correct to 3 significant figures. Hence find p.

(OCR Nov 1997 P1)

11 A, B and C are three points on horizontal ground. The distance between A and B is 285 m. Angle ABC is 73° and angle BAC is 59°. Calculate, correct to the nearest metre, the distance between A and C.

(OCR Feb 1997 P1)

12 Sketch, on a single diagram, the line L whose equation is $y = \dfrac{1}{2}$ and the curve C whose equation is $y = \sin x$, for $0 \leqslant x \leqslant 4\pi$.

Write down the exact co-ordinates of the points of intersection of L and C that are shown on your sketch.

13 **a)** Show that the equation $2 \sin^2 \theta = 1 + \cos \theta$ may be written as $2 \cos^2 \theta + \cos \theta - 1 = 0$.

   b) Hence solve the equation $2 \sin^2 \theta = 1 + \cos \theta$, giving all values of θ such that $-180° \leqslant \theta \leqslant 180°$.

14 The triangle PQR has angle PQR $= 30°$, QR $= 3$ cm, PR $= \sqrt{3}$ cm and PQ $= x$ cm.

Show that the possible values of x are given by the equation $x^2 - (3\sqrt{3})x + 6 = 0$.

Hence find the possible values of x, giving your answers in the form $k\sqrt{3}$, where k is an integer.

(OCR Nov 1997 P1)

15 At 0900 hours, a ship is at a point P which is on a bearing of $073°$ from a lighthouse L.

The distance from L to P is 22 km. The ship is sailing at a steady speed of 24 km/hr on a course bearing $203°$ and, at 1015 hours, is at the point T (see diagram).

Calculate the distance and bearing of T from L.

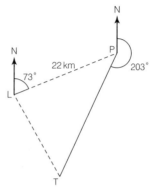

(OCR Mar 1998 P1)

16 Solve the equations

   a) $6 \sin^2 \theta = 1 + 5 \sin \theta$ $0° \leqslant \theta \leqslant 360°$

   b) $3 \sin \theta + \cos \theta = 0$ $0° \leqslant \theta \leqslant 360°$

17 In the diagram, ABC is an arc of a circle with centre O and radius 12 cm.

Angle AOC $= 1.8$ radians. Calculate the perimeter and area of the shaded region.

18 Prove that $(1 + 2 \sin \theta)^2 + (1 + 2 \cos \theta)^2 \equiv 6 + 4 \sin \theta + 4 \cos \theta$.

19 The triangle ABC has AB $= 8$ cm, AC $= 12$ cm and angle BAC $= 0.6$ radians.

   i) Calculate the area of the triangle, giving your answer correct to three significant figures.

   ii) Find the length of BC, giving your answer correct to three significant figures.

20 Solve the equation $10 \sin^2 \theta - \cos \theta - 7 = 0$ giving all values of θ such that $0 \leqslant \theta \leqslant 2\pi$. Give your answers correct to two decimal places.

21 In the diagram, ABC is an arc of a circle with centre O and radius 5 cm. The lines AD and CD are tangents to the circle at A and C respectively.

Angle AOC $= \dfrac{2}{3}\pi$ radians.

Calculate the area of the region enclosed by AD, DC and the arc ABC, giving your answer correct to 2 significant figures.

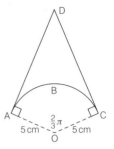

(OCR Jun 1998 P1)

22 Sketch on separate diagrams the graphs of

  a) $y = \sin x$ $0° \leqslant x \leqslant 720°$

  b) $y = 2 + \sin x$ $0° \leqslant x \leqslant 720°$

  c) $y = 2 \sin x$ $0° \leqslant x \leqslant 720°$

  d) $y = \sin 2x$ $0° \leqslant x \leqslant 720°$

23 Solve the equation $4 \sin^3 x - \sin x = 0$ giving all values of x such that $0° \leqslant x \leqslant 360°$.

Revise chapters 4, 7 and 9 before attempting this exercise.

1 Given the arithmetic progression 5.0, 5.3, 5.6, ... find

  i) the 200th term;

  ii) the sum of the first 200 terms.

(OCR Nov 1999 P2)

2 The twelfth term of an arithmetic progression is 17 and the sum of the first twelve terms is 105. Find the first term.

(OCR Mar 1996 P2)

3 **i)** The nth term of a sequence $u_1, u_2, u_3, ...$ is defined by the formula $u_n = 5 + 2 \times (-1)^n$. Write down the first six terms of the sequence.

  ii) The nth term of the oscillating sequence 7, 13, 7, 13, 7, 13, ... is denoted by v_n.

  a) Write down a formula for v_n.

  b) Evaluate $\sum_{i=1}^{201} v_i$.

4 **a)** Find the expansion of $(2x + 3y)^4$.

  b) Using your expansion with suitable values for x and y, obtain the exact value of 200.3^4.

5 A geometric progression has first term 20 and common ratio 0.9. The sum to infinity is denoted by S and the sum of the first n terms is denoted by S_n.

  a) Write down the value of S.

  b) Prove that $S - S_n = 200 \times 0.9^n$.

  c) Find the least value of n for which $S - S_n < 0.01$.

6 An arithmetic progression has first term a and common difference 3. The Nth term is 128 and the sum of the first $2N$ terms is 9842. Find N and a.

(OCR Nov 1998 P2)

7 **a)** Find the value of $\sum_{r=1}^{100} (3r - 2)$.

  b) Find the value of $\sum_{k=1}^{20} 5 \times 1.2^{k-1}$, giving your answer correct to one decimal place.

8 The nth term of a sequence is defined by $t_n = \frac{1}{2} n(n + 1)$ for all positive integers n.

  i) Find the value of $t_1 + t_2$ and of $t_2 + t_3$.

  ii) By simplifying an expression for $t_n + t_{n+1}$, show that the sum of any two consecutive terms is a perfect square.

(OCR Mar 1996 P2)

9 A geometric progression has first term 10 and second term 13. State the value of the common ratio and find, correct to the nearest integer, the 25th term.

(OCR Nov 1998 P2)

10 The sequence $u_1, u_2, u_3, \ldots$ is defined by $u_1 = 1$, $u_{n+1} = 2.5u_n$.

i) Show that $u_n = (2.5)^{n-1}$.

ii) Find the smallest integer n such that $u_n > 2^{5000}$.

iii) Find the smallest value K such that $\displaystyle\sum_{i=1}^{K} u_i > 3^{100}$.

11 An agricultural research institute is to test the effectiveness of a new liquid fertiliser in each of two fields. It has 3000 litres of fertiliser available for each field. Each field is to be divided up into a number of plots of equal size and each plot is to be treated with fertiliser.

i) The first field is to be divided into 25 plots and, from each plot to the next, the amount of fertiliser used is to increase by 3.75 litres (so that the amount used on plot 2 is 3.75 litres more than the amount used on plot 1, the amount used on plot 3 is 3.75 litres more than the amount used on plot 2, etc.). Assuming that all of the available 3000 litres of the fertiliser are used, determine the amount of fertiliser to be used on plot 1.

ii) The second field is to be divided up into N plots. The researcher decides to use 60 litres on plot 1, and to increase the amount used by 7 litres from each plot to the next. Show that $7N^2 + 113N - 6000 \leqslant 0$.

Assuming that as much as possible of the available 3000 litres of fertiliser is used, deduce the value of N and determine how much of the fertiliser will not be used.

(OCR Mar 1999 P2)

12 The first three terms in the expansion of $(1 + 3x)^6$ in ascending powers of x are the same as the first three terms in the expansion of $(1 + px + qx^2)^2$ in ascending powers of x. Find the values of p and q.

13 The sequence $u_1, u_2, u_3, \ldots$, where u_1 is a given real number, is defined for $n \geqslant 1$ by

$$u_{n+1} = \frac{2}{3}u_n + 5.$$

i) Show that

$$u_4 = \left(\frac{2}{3}\right)^3 u_1 + 5\left[1 + \frac{2}{3} + \left(\frac{2}{3}\right)^2\right]$$

and find a similar expression for u_5 in terms of u_1.

ii) Write down a similar expression for u_n in terms of u_1 and n.

iii) Show that the sequence converges to a limit. Determine this limit, showing that it is independent of u_1.

(OCR Nov 1998 P2)

14 The first three terms of an arithmetic progression are 2, 7, 12. Find an expression in terms of n for u_n, the nth term of this arithmetic progression. A second arithmetic progression is such that its nth term, w_n, is given by $w_n = 3n + 67$.

   i) Show that $u_{35} = w_{35}$.

   ii) Show that $\sum_{n=1}^{35} (w_n - u_n) = 1190$.

   iii) Find the value of p for which $\sum_{n=1}^{p} (u_n - w_n) = 370$.

(OCR Nov 1997 P2)

15 A geometric progression has common ratio r and its second term is -10. Given that the sum to infinity is 9, show that $9r^2 - 9r - 10 = 0$. Find the value of the common ratio.

(OCR Jun 1997 P2)

16   **a)** Find the first four terms in the expansion of $(2 + 3x)^7$.

   b) Write down the first four terms in the expansion of $(2 - 3x)^7$.

   c) Without using a calculator and showing your method clearly, estimate the value of $2.06^7 - 1.94^7$.

17 The sequence u_1, u_2, u_3, ..., where u_1 is a given real number, is defined for $n \geqslant 1$ by

$$u_{n+1} = 1 - \frac{1}{u_n}.$$

   i) For the particular case when $u_1 = 4$, find u_2, u_3 and u_4, and describe the behaviour of the sequence.

   ii) For the general case when $u_1 = k$ (where k is not equal to 0 or 1)

      a) show that $u_3 = -\dfrac{1}{k-1}$ and deduce that $u_4 = k$,

      b) show that $u_1 u_2 u_3 = -1$,

      c) state the value of $u_{100} u_{101} u_{102}$.

   iii) Given that $u_1 = 2$, evaluate $\sum_{i=1}^{1000} (u_i)^2$.

(OCR Mar 1999 P2)

18   **i)** Determine the range of values of x for which the geometric series $1 + 4^x + 4^{2x} + 4^{3x} + \cdots$ is convergent (i.e. has a sum to infinity).

   ii) Given that the sum to infinity of this series is 10, prove that $x = \dfrac{\log 0.9}{\log 4}$.

19 Find the first three terms in the expansion of $(1 - 2x)^5$.

In the expansion of $(a + bx)(1 - 2x)^5$ the coefficient of x is -17 and the coefficient of x^2 is 50. Determine the values of a and b.

20 The sequence u_1, u_2, u_3, ..., where u_1 is a given real number, is defined for $n \geqslant 1$ by

$$u_{n+1} = \frac{2}{u_n}.$$

    i) For the case when $u_1 = -1$, write down the first four terms of the sequence and hence describe its behaviour.

    ii) Find the two values of u_1 for which the terms of the sequence are all the same.

(OCR Mar 1998 P2)

21 A post is being driven into the ground by a mechanical hammer. The distance it is driven by the first blow is 8 cm. Subsequently, the distance it is driven by each blow is $\frac{9}{10}$ of the distance it was driven on the previous blow.

    i) The post is to be driven a total distance of at least 70 cm into the ground. Find the smallest number of blows needed.

    ii) Explain why the post can never be driven a total distance of more than 80 cm into the ground.

(OCR Mar 1997 P2)

22 The sum of the first two terms of an arithmetic progression is 18 and the sum of the first four terms is 52. Find the sum of the first eight terms.

(OCR Mar 1997 P2)

23 The sequence u_1, u_2, u_3, ... is defined by $u_n = 2n^2$.

    i) Write down the value of u_3.

    ii) Express $u_{n+1} - u_n$ in terms of n, simplifying your answer.

    iii) The differences between successive terms of the sequence form an arithmetic progression. For this arithmetic progression, state its first term and its common difference, and find the sum of its first 1000 terms.

(OCR Jun 1995 P2)

24   **i)** The eighth term of an arithmetic progression is 1 and the fifteenth term is 4.5. Show that the sum of the first 95 terms is 1995.

    ii) A second arithmetic progression has first term 6 and common difference 3. The sum of the first n terms of this arithmetic progression is also 1995. Find the value of n.

(OCR Nov 1995 P2)

Revise chapters 3 and 6 before attempting this exercise.

1   Given that $(x + 2)$ is a factor of $ax^3 + 2x^2 - ax + 10$ find the value of the constant a.

(OCR Mar 1998 P1)

2   **a)**   Solve the equation $\log_3\left(\dfrac{1}{2}x + 3\right) = 2$.

       b)   Solve the equation $3^{2t+1} = 100\,000$, giving your answer correct to three significant figures.

3   The value of a certain make of car when new is £13 500 and the value depreciates by 15% each year. Hence the value when the car is one year old is £13 500 × 0.85. Find the value of this make of car when it is eight years old, giving your answer to the nearest pound.

    The value of a different make of car depreciates by 18% each year. Its value when new is £19 620. After n years its value is £1000. Write down an equation satisfied by n, and hence find n.

(OCR Nov 1997 P2)

4   Given that the equation $6x^3 - 19x^2 + x + 6 = 0$ has an integer root, find all the roots of the equation.

5   Sketch the graphs of

    a) $y = 20 \times 1.2^x$ **b)** $y = 400 \times 0.92^x$

    taking care to mark clearly where the curves cross the y axis.

6   Find the remainder when $2x^4 + 7x^3 + 3x - 2$ is divided by $(x + 1)$.

7   The mass, m grams, of a radioactive substance after t days (where $t \geqslant 0$) is given by $m = 720a^t$ where a is a constant. The mass decreases as time increases.

    i)   What can be said about the constant a?

    ii)   Sketch a graph showing how the mass varies with time.

    After 60 days the mass will have decreased to 80% of its value when $t = 0$.

    iii) Determine the value of a, giving your answer correct to four significant figures.

    iv) Find the time required for the mass to decrease to half of its original value.

8   Find constants P, Q, R and S such that $\dfrac{x^3 + 5x^2 + x - 4}{x + 2} \equiv Px^2 + Qx + R + \dfrac{S}{x + 2}$.

9   **a)**   Write down the values of   **i)** $\log_4 64$ **ii)** $\log_{25}\left(\dfrac{1}{5}\right)$.

       b)   Find the value of $6\log_{12} 2 + 2\log_{12} 15 - 2\log_{12} 10$.

       c)   Solve the equation $\log_7(5x - 1) = 2$.

10 Solve the inequalities   **a)** $2.3^x > 150$ **b)** $500 \times 0.8^x - 20 > 35$.

11 The polynomial p(x) is defined by $p(x) = 2x^3 + ax^2 - 21x + b$, where a and b are constants. p(x) has a factor of $(x + 3)$ and has a remainder of 20 when divided by $(x - 2)$.
  a) Determine the values of the constants a and b.
  b) Solve the equation $p(x) = 0$.
  c) Divide p(x) by $(x - 4)$.

12 Given that $\dfrac{3}{5}$ is a root of the equation $50x^3 + px^2 - 18x - 9 = 0$
  i) determine the value of the constant p,
  ii) find the other two roots of the equation $50x^3 + px^2 - 18x - 9 = 0$.

13 Find the value of 2006^{2005}, giving your answer in standard form, correct to three significant figures.

14 Given that $p = 2^{300}$ and $q = \dfrac{1}{2^{50}}$
  a) write down the values of $\log_2 p$ and $\log_2 q$,
  b) find the value of $\log_2(p^2 q)$,
  c) find the value of $\log_2\left(\dfrac{\sqrt{p}}{q^2}\right)$,
  d) find the value of x if $\log_4(qx) = -3$, giving your answer as a power of 2.

15 By putting $t = 2^x$, show that the equation $2^{2x+3} - 39 \times 2^x + 45 = 0$ may be written as $8t^2 - 39t + 45 = 0$. Hence solve the equation $2^{2x+3} - 39 \times 2^x + 45 = 0$, giving your answer correct to three decimal places.

16 Obtain the remainder when $8x^4 + 4x^2 + 6x - 3$
  a) is divided by $(x - 2)$,
  b) is divided by $(2x - 1)$.

17 a) If $3^{y+2} = 7^x$ prove that $y = kx - 2$ where $k = \dfrac{\log 7}{\log 3}$.
  b) If $\log_2(x^5) - \log_2(xy^2) + 3 \log_2(y) = 5$, prove that $y = kx^n$, giving the values of the constants k and n.

18 A population of bacteria is growing exponentially. Explain what this means using a sketch graph to illustrate your answer.

At time t hours, the size s of a population of bacteria is given by $s = 500 \times 1.04^t$.
  a) State the value of s when $t = 0$.
  b) Find the size of the population when $t = 10$.
  c) Find the value of t when the size of the population is three times its initial value, giving your answer correct to 3 significant figures.

19 a) Find the quotient and remainder when
$x^2 + 7x - 8$ is divided by $x - 3$.

The diagram shows the graph of
$$y = \frac{x^2 + 7x - 8}{x - 3}.$$

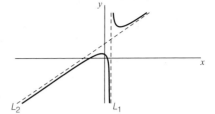

b) Write down the equation of the vertical line, L_1, that the curve does not cross.

c) Use your answer to (a) to write down the equation of the line L_2, that the line approaches as $x \longrightarrow \pm \infty$.

20 The cubic polynomial $x^3 - 2x^2 - 2x + 4$ has a factor $(x - a)$ where a is an integer.

i) Use the factor theorem to find the value of a.

ii) Hence find exactly all three of the roots of the cubic equation $x^3 - 2x^2 - 2x + 4 = 0$.

21 Solve the equation $\log_4(5x + 1) - \log_4(x + 2) = 1$.

22 Solve the equation $x^3 - 6x^2 + 7x + 2 = 0$ given that one of the roots is an integer.

23 a) The cubic polynomial $x^3 - kx^2 - 5x + 30$, where k is a constant, has a factor $(x - k)$. Find the value of k.

b) Find a cubic equation with integer coefficients whose roots are 5, -3 and $\dfrac{7}{3}$.

24 If $7^{x+2} = 5^{2x+1}$ prove that $x = \dfrac{\log\left(\dfrac{49}{5}\right)}{\log\left(\dfrac{25}{7}\right)}$.

25 Solve the equation $\log_5(24x + 4) - 2\log_5(x - 2) = 2$.

Revise chapter 5 before attempting this exercise.

1 Find   **i)** $\int x(x+1)\,dx$ **ii)** $\int \frac{1}{x^2}\,dx$

(OCR Jan 2001 P1)

2 Sketch the graph of $y = 3\sqrt{x}$. Find the area of the closed region formed by

   i) the curve $y = 3\sqrt{x}$, the x axis and the lines $x = 1$ and $x = 4$,

   ii) the curve $y = 3\sqrt{x}$, the y axis and the lines $y = 6$ and $y = 9$.

3 The diagram shows the curve $y = 8 - 2x^2$ together with the straight line $y = 5 - x$.

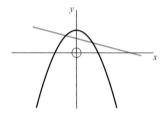

   a) Find the co-ordinates of the points where the line and the curve intersect.

   b) Show that the area of the closed region enclosed by the line and the curve is $\frac{125}{24}$.

(OCR Jan 2001 P1)

4 A curve passes through the point (2, 5) and has gradient given by $\frac{dy}{dx} = 6x^2 - \frac{8}{x^3}$.

   Find the equation of the curve.

5 **a)** Evaluate $\int_1^4 x^2 + 4x\,dx$.

   b) Find the value of a if $\int_1^a \frac{18}{x^3}\,dx = 8$.

6 Sketch the graph of $y = 36 - 4x^2$, giving the x co-ordinates of each of the points where the curve crosses the x axis.

   Find by integration the area of the closed region bounded by the x axis and the curve.

7 Evaluate $\int_0^8 \frac{1}{\sqrt[3]{y}}\,dy$.

8 The diagram shows the graph of $y = g(x)$.

   It is given that $\int_0^8 g(x)\,dx = 10$ and $\int_3^8 g(x)\,dx = 16$.

   Find the value of $\int_0^3 g(x)\,dx$.

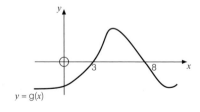

$y = g(x)$

9 The diagram shows the graph of $y = x^{\frac{3}{2}} - 3x$. The curve crosses the positive x axis at P.

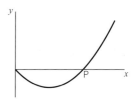

    a) Calculate the co-ordinates of the point P.

    b) Prove that the closed region bounded by the curve and the x axis has area 24.3.

10 Evaluate

    a) $\displaystyle\int_{1}^{2} (3x-1)^2 \, dx$ **b)** $\displaystyle\int_{1}^{2} \frac{x^2+8}{x^5} \, dx$

11 The diagram shows the curve $y = \sqrt{x}$ and the normal to the curve at the point P(4, 2).

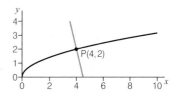

    a) Prove that the normal has equation $y = -4x + 18$.

    b) Prove that the closed region bounded by the curve $y = \sqrt{x}$, the normal and the x axis has area $\dfrac{35}{6}$.

12 Find the area of the closed region formed by the curve $y = x^3$, the y axis and the lines $y = 1$ and $y = 8$.

13 Evaluate   **a)** $\displaystyle\int_{2}^{\infty} \frac{48}{x^3} \, dx$ **b)** $\displaystyle\int_{0}^{4} \frac{3}{\sqrt{x}} \, dx$

14 Find the points of intersection of the curves $y = 5x^3$ and $y = x^4$. Calculate the area of the region bounded by these two curves.

15 **a)** Find $\displaystyle\int x^2 + \frac{4}{x^2} \, dx$ **b)** Evaluate $\displaystyle\int_{0}^{4} \frac{t^2+4}{\sqrt{t}} \, dt$

16 The diagram shows a sketch of the graph of $y = 2^{4-x^2}$. The table shows values of y for x – 0, 0.5, 1, 1.5, 2, 2.5 and 3.

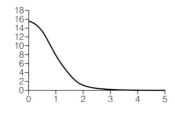

x	y
0	16
0.5	13.454
1	8
1.5	3.364
2	1
2.5	0.21
3	0.031

Use the trapezium rule with seven ordinates to estimate the value of $\displaystyle\int_{0}^{3} 2^{4-x^2} \, dx$.

17 Find the equation of the curve passing through (1, 5) for which $\dfrac{dy}{dx} = 6x + 4$.

The diagram shows the graph of this curve. Find the area of the shaded region.

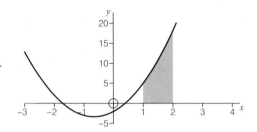

18 The diagram shows the graph of $y = \sqrt{4x - 12}$.

Calculate the area of the closed region bounded by the curve, the y axis and the lines $y = 2$ and $y = 4$.

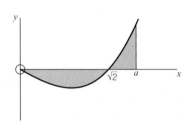

19 Find $\displaystyle\int x(x^2 - 2)\,dx$.

The diagram shows the graph of $y = x(x^2 - 2)$ for $x \geqslant 0$. The value of a is such that the two shaded regions have equal areas. Find the value of a.

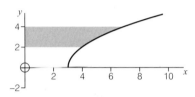

(OCR Mar 1997 P2)

20 The diagram shows the graph of

$$y = 2\sqrt{1 - \dfrac{x^2}{9}} \text{ for positive values of } x.$$

Use the trapezium rule with three strips to estimate the value of

$$\int_0^3 2\sqrt{1 - \dfrac{x^2}{9}}\,dx$$

and state, with a reason, whether your estimate is an under-estimate or an over-estimate of the correct value.

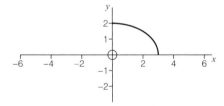

Hence estimate the area of the interior of the ellipse $\dfrac{x^2}{9} + \dfrac{y^2}{4} = 1$ which is shown in the second diagram.

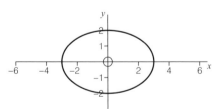

21 The diagram shows the curve $y = x^2$ and the normal to the curve at the point $(1, 1)$.

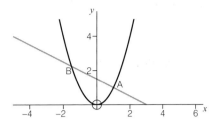

a) Find the equation of the normal to the curve $y = x^2$ at the point A$(1, 1)$. Verify that the point B where the normal cuts the curve again has co-ordinates $\left(-\dfrac{3}{2}, \dfrac{9}{4}\right)$.

b) Calculate the area of the closed region which is bounded by the curve and the normal, leaving your answer as an exact fraction.

(OCR Mar 1996 P2)

22 Figure 1 shows part of the curve $y = \dfrac{1}{2}x^2 - \dfrac{1}{20}x^4$. Show that the two maximum points on the curve occur when $x = \pm\sqrt{5}$.

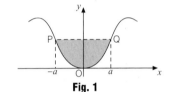

Fig. 1

The points P and Q on the curve have x co-ordinates $-a$ and a, respectively, where a is a positive constant such that $a < \sqrt{5}$. The shaded region is bounded by the curve and by the line segment PQ.
Show that the area of this shaded region is

$$\frac{2}{75}a^3(25 - 3a^2).$$

This curve forms the cross-section of a straight, horizontal drainage channel, as shown in Figure 2. The units involved are metres.

Fig. 2

i) After heavy rain, the channel is full to overflowing. Determine the greatest depth of the water in the channel.

ii) On another occasion, the water in the channel is such that its greatest depth is 0.45 m. Find the volume of water in a 30-metre length of the channel.

(OCR Nov 1999 P2)

1 Find the binomial expansion of $(2 + x)^4$, simplifying the terms. [3]

2 Calculate the exact area of the shaded region between the two curves.

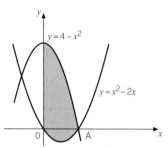

[6]

3 **i)** Show that the sum of all the positive integers from N to $2N$ inclusive is

$$\frac{3}{2}N(N + 1).$$ [4]

ii) Hence, or otherwise, find the sum of all the multiples of 5 from 500 to 1000 inclusive. [3]

4 The diagram shows a circle with centre O and radius 6 cm. The chord AB has length 10 cm and the angle AOB is θ radians. Find

i) the value of θ, [3]

ii) the area of the shaded segment. [5]

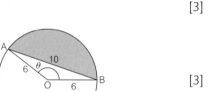

5 **a)** Find $\displaystyle\int_4^9 \frac{3}{\sqrt{x}}\,dx$. [4]

b) It is given that $\displaystyle\int_2^a 1 + \frac{1}{x^2}\,dx = 2\frac{1}{4}$, where a is an integer. Find the value of a. [6]

6 The cubic polynomial $f(x) = 4x^3 - kx^2 + (k - 3)x - 3$ has a factor $x - 3$.

i) Find the value of k. [2]

ii) Factorise $f(x)$ completely and hence solve the equation $f(x) = 0$. [6]

7 **a)** Find the exact value of $3 \tan 60° + 4 \cos 30°$. [2]

b) Given that θ is an acute angle, simplify $\dfrac{\sqrt{1 - \cos^2 \theta}}{\cos \theta}$, expressing your answer in terms of a single trigonometric function. [2]

c) Solve the equation

$$3 \sin \theta \cos \theta = 2 \sin \theta$$

for $0° \leqslant \theta \leqslant 180°$, giving your answers correct to 1 decimal place where necessary. [5]

8 The diagram shows the curve $y = \left(\frac{2}{3}\right)^x$, together with a series of rectangles of unit width drawn beneath the curve.

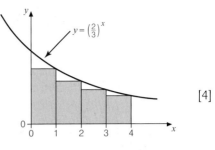

i) Given that the rectangle between $x = n$ and $x = n + 1$ is the first to have an area less than 0.001 $unit^2$, find the value of n, [4]

ii) Show that the areas of successive rectangles form a geometric progression and that the sum of all the rectangles from $x = 0$ to $x = 100$ can be written in the form

$$p\left[1 - \left(\frac{2}{3}\right)^q\right]$$

where p and q are integers, whose values should be found. [5]

9 **a)** Given that $y = a(1 - 2^{kx})$, express x in terms of a k and y. [4]

b) Solve the equation $\log_{10}(x^2 + 15x) = 2$ [4]

c) Simplify

$$\log_n \sqrt{1 + n^2} - \frac{1}{2}\log_n\left(1 + \frac{1}{n^2}\right),$$

showing that the result is independent of n. [4]

ANSWERS

CHAPTER 1
Exercise 1

1 **a)** 1, 0   **b)** 0, 1   **c)** −1, 0
  d) 0, −1   **e)** 1, 0   **f)** 0, −1
  g) −1, 0   **h)** 0, 1

2 **a)** 0.87, −0.5   **b)** −0.87, 0.5
  c) −0.87, −0.5   **d)** 0.87, −0.5
  e) 0.87, 0.5   **f)** −0.87, 0.5
  g) −0.87, −0.5   **h)** 0.87, −0.5
  i) −0.87, −0.5   **j)** −0.87, 0.5
  k) 0.87, 0.5

3 **a)** 0.5, −0.87   **b)** −0.5, 0.87
  c) −0.5, −0.87   **d)** 0.5, −0.87
  e) 0.5, 0.87   **f)** −0.5, 0.87
  g) −0.5, −0.87   **h)** 0.5, −0.87
  i) −0.5, −0.87   **j)** −0.5, 0.87
  k) 0.5, 0.87

CHAPTER 1
Exercise 2

4 **a)** $y = 3 \sin \theta$ **b)** $y = \cos 4\theta$
  c) $y = \cos \theta - 2$

CHAPTER 1
Exercise 3

1 **a)** $\dfrac{1}{2}$ **b)** $\dfrac{-\sqrt{3}}{2}$ **c)** $\dfrac{-\sqrt{3}}{3}$
  d) $-\dfrac{1}{2}$ **e)** $\dfrac{-\sqrt{3}}{2}$ **f)** $\dfrac{\sqrt{3}}{3}$
  g) $\dfrac{-\sqrt{3}}{2}$ **h)** $\dfrac{-1}{2}$ **i)** $\sqrt{3}$
  j) $\dfrac{-\sqrt{3}}{2}$ **k)** $\dfrac{1}{2}$ **l)** $-\sqrt{3}$
  m) $\dfrac{-1}{2}$ **n)** $\dfrac{\sqrt{3}}{2}$ **o)** $\dfrac{-\sqrt{3}}{3}$

2 **a)** $\dfrac{\sqrt{2}}{2}$ **b)** $\dfrac{-\sqrt{2}}{2}$ **c)** −1
  d) $\dfrac{-\sqrt{2}}{2}$ **e)** $\dfrac{-\sqrt{2}}{2}$ **f)** 1
  g) $\dfrac{-\sqrt{2}}{2}$ **h)** $\dfrac{\sqrt{2}}{2}$ **i)** −1

3 **a)** 1   **b)** 1   **c)** 1

4 **a)** $\sqrt{3}$ **b)** $\dfrac{\sqrt{3}}{3}$ **c)** 0

CHAPTER 1
Exercise 4

1 $\dfrac{5}{13}$ $\dfrac{5}{12}$

2 $\dfrac{-16}{65}$ $\dfrac{16}{63}$

3 $\dfrac{-\sqrt{5}}{3}$ $\dfrac{-2\sqrt{5}}{5}$

5 $\sin 2\theta \equiv 2 \sin \theta \cos \theta$

6 $\cos 2\theta \equiv \cos^2 \theta - \sin^2 \theta$

7 $\tan 2\theta \equiv \dfrac{2 \tan \theta}{1 - \tan^2 \theta}$

8 $a = 3$

CHAPTER 1
Exercise 5

1 **a)** −300, −60, 60, 300, 420, 660
  b) 135, 225
  c) 76.5, 283.5, 436.5, 643.5
  d) −300, −240, 60, 120
  e) −120, −60, 240, 300
  f) −720, −540, −360, −180, 0
  g) −225, −45, 135, 315
  h) 71.6, 251.6, 431.6, 611.6
  i) No solutions

2 **a)** −180, −60, 60, 180
  b) 0, 90, 180, 270, 360
  c) 22.5, 67.5, 112.5, 157.5
  d) −110, −70, 10, 50, 130, 170
  e) 15, 75, 135
  f) −84, −48, −12, 24, 60
  g) 105, 165, 285, 345

3 $t = \frac{2}{3}, 3\frac{1}{3}, 4\frac{2}{3}, 7\frac{1}{3}$

CHAPTER 1
Exercise 6

1 45, 135, 225, 315

2 0, 90, 180, 360

3 0, 360, 131.8, 228.2

174

4 $\theta = 0, 45, 180, 315, 360$

5 $\theta = 90, 210, 330$

6 $\theta = 30, 150, 199.5, 340.5$

7 $\theta = -180, -60, 60, 180$

8 $\theta = -51.3°, -128.7°$

9 $\theta = -150, -30, 30, 150$

10 $-180, -120, 0, 120, 180$

11 a) BC = 15 cos θ BP = 15 cos$^2 \theta$
       DF = 5 sin θ
    c) $\theta = 57.9$ Area = 17.94

CHAPTER 1
Revision Exercise

1 $x = 150, 270, 510, 630$

2 b) $\theta = 0, 60, 300, 360$

3 $\sin \alpha = \dfrac{\sqrt{15}}{8}$ $\tan \alpha = \dfrac{\sqrt{15}}{7}$

4 b) $\theta = 30, 150, 221.8, 318.2$

5 a) Stretch factor $\frac{1}{5}$ in x-direction.
       Period = 72°
    b) $x = 12, 24, 84$

6 b) 2 solutions   **c)** $a = 3$ $b = 1$ $c = -3$
    d) $x = 57.9°, 122.1°$

7 b) $\theta = 9.2°, 99.2°, 189.2°, 279.2°$

8 a) $\cos \alpha = \dfrac{\sqrt{5}}{5}$ $\sin \alpha = \dfrac{2\sqrt{5}}{5}$

    b) $\cos \beta = \dfrac{-\sqrt{5}}{5}$ $\tan \beta = \dfrac{-2\sqrt{5}}{5}$

CHAPTER 2
Exercise 1

1 12.57 cm

2 27.86°

3 15.09 cm

4 98.25°

5 11.36 km

6 a) 3577 m   **b)** 8813 m   **c)** 12 150 m

7 $\theta = 118.07°$

8 $\theta = 21.67°$

CHAPTER 2
Exercise 2

1 a) 96.3 cm^2 **b)** 220.9 cm^2
    c) 40.1 cm^2

2 27 817 m^2

CHAPTER 2
Exercise 3

1 $x = 15.8$ cm $y = 16.1$ cm

2 $\theta = 30.2°$

3 $x = 10.5$ cm $y = 16.8$ cm

4 $x = 17.7$ cm $\theta = 40.3°$

5 $x = 7.29$ cm $\theta = 154.8°$

6 $\theta = 138.3°$

7 8.12 cm   15.3 cm

8 $\angle C_1 = 27.40°$ $\angle C_2 = 152.60°$
    BC_1 11.7 cm BC_2 3.7 cm

CHAPTER 2
Revision Exercise

1 $\theta = 57.45°$ $x = 7.95$ cm

2 $y = 5.84$ cm $\theta = 115.6°$

3 $x = 5.48$ cm $y = 10.30$ cm

4 AC = 2266 m   bearing = 295°

5 BD = 13 cm $\theta = 32.2°$
    Area $\approx$ 103.9 cm^2

6 $\theta_1 = 56.4°$ $\theta_2 = 123.6°$
    MQ = 11.98 miles   MP = 5.34 miles
          11.1 km/hr

8 i) $2\sqrt{14}$ **ii)** $6\sqrt{3}$

9 $\theta = 37.28°$ Area = 8.40 cm^2

CHAPTER 3
Exercise 1

1 $x^2 + 2x + 2 + \frac{-3}{x-1}$

2 $5x - 12 + \frac{41}{x+4}$

3 $x^2 + 2x + 7 + \frac{10}{x-2}$

4 $5 + \frac{-17}{x+4}$

5 $2x - 6 + \frac{18}{x+3}$

6 $2 + \frac{-13}{x+3}$

7 $5x - 2 + \frac{8}{2x+3}$

8 $2x^2 + x - 3 + \frac{5}{3x-1}$

9 $5x + 15 + \frac{49}{x-3}$

10 $x^2 - \frac{1}{2}x - \frac{7}{4} + \dfrac{-\frac{13}{4}}{2x+1}$

11 $x^2 + 1 + \frac{7}{x-2}$

12 $3x^2 + x + 1$

13 a) $x + 7 + \frac{12}{x-2}$ **b)** 12 **c)** 12

14 a) $3x + 2 + \frac{-20}{x-3}$ **b)** −20 **c)** −20

15 a) $\dfrac{3x^3 + 7}{x+1} = 3x^2 - 3x + 3 + \dfrac{4}{x+1}$
 b) 4
 c) 4

CHAPTER 3
Exercise 2

1 a) $2x^2(x-2)(x-3)$
 b) $5t^3(t-6)(t-10)$

2 a) $(x-5)(x+5)(x^2+4)$
 b) $(2x-3)(2x+3)(x-2)(x+2)$

3 $p = 3, q = 7, r = -26$
 $(x-3)(3x+13)(x-2)$

4 $(x-3)(2x+5)(x+7)$

5 $(x+7)(4x-1)(x+2)$

6 $4x(x-3)(x+1)$

7 $(x-3)(x+3)(x-1)(x+1)$

8 $y^2(y-1)(y+1)(y-2)(y+2)$

9 $(3t^2+7)(t-2)(t+2)$

10 $x^2(3x-10)(x+1)$

11 $(y-3)(y+3)(y^2+9)$

12 $(y^2+5)(y+2)(y-2)$

13 $x^2(2x-3)(x+1)$

14 $(2x^2-3)(x^2+1)$

15 $(2y-3)(2y+3)(y-1)(y+1)$

CHAPTER 3
Exercise 3

1 $x = 2, 6, -1$

2 $(y+3)(y^2+6)$

3 $a = 6$

4 i) $a = -14$ $b = -36$
 ii) $x = 1, 4$ or 9
 iii) $y = \pm1, \pm2, \pm3$

5 a) $x = 3$ or $\dfrac{-5 \pm \sqrt{33}}{2}$
 b) $x = 4$ $1 \pm 2\sqrt{2}$

6 $8x^2 + 2x - 3 = 0$

7 a) 8 **b)** 48 **c)** 982

8 i) $a = -5$ $b = -4$
 ii) $-1, \frac{1}{2}, 3$

9 a) 5 **b)** −4

10 a) $m = -3$ **b)** −4

11 $a = 7, b = -54$

12 $5x^3 + 2x^2 - 33x + 18 = 0$

CHAPTER 3
Extension Exercise

1 $x - 1 + \dfrac{6x+2}{x^2+x+2}$

2 $3x^2 + 3 + \dfrac{3}{x^2-1}$

CHAPTER 3
Revision Exercise

1 $a = -5, 2$

2 a) $b = 6$ **b)** 160

3 a) 24
 b) $x = -2$ or $\dfrac{-1 \pm \sqrt{33}}{4}$

4 a) Quotient $= x + 2$ Remainder $= -8$
 b) Quotient $= 2x^2 + 8x + 37$
  Remainder $= 141$
 c) Quotient $= 2x^2 + 3x + 9$
  Remainder $= 20$

5 **i)** $a = 54$ **ii)** $b = -7$

6 $(x + 1)(3x - 2)(2x - 5)$
$(x - 2)(x + 2)(x - 5)(x + 5)$

7 $a = 3$

8 **a)** $a = 8$ **b)** 182   **c)** just one root

9 **a)** -2 **b)** $x = 2$ or $3 \pm \sqrt{10}$

10 $x = -3, \frac{1}{2}, 4.$

CHAPTER 4
Exercise 1

1 12, 6, 4, 3, 2.4 $x_n \rightarrow 0$

2 $\frac{3}{2}, \frac{8}{3}, \frac{13}{4}, \frac{18}{5}, \frac{23}{6}$ $x_n \rightarrow 5$

3 $-1, 1, -1, 1, -1$
x_n oscillates between -1 and 1.

4 1, 4, 9, 16, 25 $x_n \rightarrow \infty$

5 2, 4, 6, 8, 16, 32 $x_n \rightarrow \infty$

6 1.02, 1.04, 1.06, 1.08, 1.10
(2 d.p.) $x_n \rightarrow \infty$

7 0.9, 0.81, 0.729, 0.6561,
0.59049 $x_n \rightarrow 0$

8 $-0.6, 0.36, -0.216, 0.1296,$
-0.07776 $x_n \rightarrow 0$

9 $-2.1, 4.41, -9.26, 19.45, -40.84$
(2 d.p.) $x_n \rightarrow \infty$

10 $\frac{3\sqrt{2}}{2}, 0, \frac{-3\sqrt{2}}{2}, -3, \frac{-3\sqrt{2}}{2}$
Values of x_n repeat cyclically with
period 8

11 7, 12, 17, 22, 27 $x_n \rightarrow \infty$

12 $\frac{-1}{3}, \frac{2}{4}, \frac{-3}{5}, \frac{4}{6}, \frac{-5}{7}$
x_n oscillates between 1 and -1
approximately

13 (c), (d), (e) $-1 < \alpha < 1$

14 (d), (e)
$x_n \rightarrow 0$ if $\beta < 0$
$x_n \rightarrow \infty$ if $\beta > 0$

15 **i)** $x_n \rightarrow 7$ $x_n = 7 - \frac{33}{n + 4}$
   ii) $x_n \rightarrow -2$ $x_n = \frac{22}{n + 1} - 2$

16 **a)** $h_1 = 1.6$ $h_2 = 1.28$ $h_3 = 1.024$
   $h_4 = 0.8192$ $h_5 = 0.65536$
   b) $h_n \rightarrow 0$

17 **a)** £10   **b)** £3   **c)** $n = 64$ months

CHAPTER 4
Exercise 2

1 5, 18, 25.8, 30.48, 33.288
$L = 37.5$

2 8, 28.4, 24.36, 25.136, 24.9728
$L = 25$

3 2, 11.9, 21.31, 30.24, 38.73 (2 d.p.)
$L = 200$

4 2, 4, 2.4, 3.529 $\cdots$, 2.649 $\cdots$
$L = 3$

5 4, 1, 2, 1.5, 1.714 $\cdots$
$L = -1 + \sqrt{7}$

6 $a_n \rightarrow 75$

7 Diverges

8 Diverges

9 $a_n \rightarrow 6.25$

10 $u_2 = 6, u_3 = 2, u_4 = 6, u_5 = 2$
Oscillates between 6 and 2.
$u_1 = \pm\sqrt{12}$

11 **b)** $t_5 = 91\,477$
   c) $L = 75\,000$

12 **b)** North = 27 588 000
   South = 32 412 000
   c) $L = 48\,000\,000$

CHAPTER 4
Exercise 3

Q1–7 other correct answers possible

1 $\displaystyle\sum_{r=5}^{20} r^3$

2 $\displaystyle\sum_{r=6}^{25} 2^r$

3 $\displaystyle\sum_{r=7}^{73} \frac{4}{r}$

4 $\displaystyle\sum_{r=3}^{100} \frac{r+1}{r^2}$

5 $\displaystyle\sum_{r=1}^{n} a_r^2$

6 $\displaystyle\sum_{r=1}^{n} ra_r$

7 $\displaystyle\sum_{r=1}^{n} \frac{r}{a_r}$

8 $1^3 + 2^3 + 3^3 + 4^3 + 5^3$

9 $\frac{12}{3^2} + \frac{12}{4^2} + \frac{12}{5^2} + \frac{12}{6^2}$

10 $3 \times 2^1 + 3 \times 2^2 + \cdots + 3 \times 2^{19} + 3 \times 2^{20}$

11 $a_1 + a_2 + a_3 + a_4 + a_5$

12 $\frac{5}{a_4} + \frac{6}{a_5} + \frac{7}{a_6} + \frac{8}{a_7} + \frac{9}{a_8}$

13 385

14 1.635

15 7.485

16 a) 1218   **b)** 630   **c)** 1122
    d) 504   **e)** 462

CHAPTER 4
Revision Exercise

1   **a)** 98, 89, 80.9, 73.6   **b)** $u_n \to 8$

2   86

3   **a)** $y_2 = 760;\ y_3 = 616;\ y_4 = 529.6$
    b) $L = 400$

4   **a)** $x_2 = 7.143;\ x_3 = 5.469;\ x_4 = 6.695$
    b) $M = -1 + \sqrt{51}$

5   **a)** $x_1 = \dfrac{\sqrt{3}}{2}$ $x_2 = \dfrac{\sqrt{3}}{2}$ $x_3 = 0$ $x_4 = \dfrac{\sqrt{3}}{2}$

      $x_5 = \dfrac{\sqrt{3}}{2}$ $x_3 = 0$ $x_7 = \dfrac{\sqrt{3}}{2}$ $x_8 = \dfrac{\sqrt{3}}{2} \cdots$

    b) $\dfrac{\sqrt{3}}{2}$ **c)** $\displaystyle\sum_{1}^{300} x_r = 0$

6   **a)** $x_0 = 2000$
    c) $x_1 = 1800;\ x_2 = 1650;\ x_3 = 1537.5$
    d) $L = 1200$

7   **a)** $4 + \dfrac{9}{2n-1}$
    b) $x_1 = 13$ $x_2 = 7$ $x_3 = 5.8$
    c) $x_n \to 4$
    d) 36.09

8   **i)**   **a)** $u_1 = 2$ $u_2 = 4$ $u_3 = 2$
          $u_4 = 4$ $u_5 = 2$
          Oscillates between 2 and 4

      b) $\displaystyle\sum_{1}^{5} u_i = 14$ **c)** 300

    ii) $u_1 = 3$

9   **a)** 5071   **b)** 12 100   **c)** 3065

CHAPTER 5
Exercise 1

1 $8x^3 + 5$

2 $24x^3 + 10x$

3 $\dfrac{-8}{x^3}$

4 $\dfrac{-4}{x^2} + \dfrac{9}{x^4}$

5 $6x + \dfrac{12}{x^4}$

6 $3x^{-0.5} + 2x^{-1.5}$

7 $3 - \dfrac{10}{x^3}$

8 $7.5x^{6.5}$

9 $1 - 3x^{-0.5}$

10 $2x^{-\frac{2}{3}}$

11 -3

12 -3

13   2

14   20

15 $\frac{11}{3}$

CHAPTER 5
Exercise 2

1 $y = 2x^2 + 3x + c$

2 $y = x^3 + 3x^2 - 2x + c$

3 $y = \dfrac{4}{x} + c$

4 $y = 2x^4 - 3x^2 + c$

5 $y = \frac{1}{3}x^3 + \frac{3}{2}x^2 + x + c$

6 $y = 2x^5 - 3x^4 + 4x^2 - 3x + c$

7 $y = \dfrac{-10}{x} + c$

8 $y = \dfrac{-5}{x^2} + c$

9 $y = 3x^2 - 11x + 16$

10 $y = 4x^3 - 4x^2 + x + 11$

11 $3x^3 - 4x^2 + 5x + c$

12 $\frac{3}{2}x^2 + 2x + c$

13 $\frac{7}{2}x^2 - 3x + c$

14 $\dfrac{-3}{x^3} + c$

15 $2x^3 + \dfrac{6}{x^2} + c$

CHAPTER 5
Exercise 3

1 $\frac{5}{3}x^3 + c$

2 $2u^4 + c$

3 $4s^{1.5} + c$

4 $\dfrac{-12}{x} + c$

5 $2x^{0.5} + c$

6 $\frac{5}{6}x^{\frac{6}{5}} + c$

7 $\frac{5}{7}t^{\frac{7}{5}} + c$

8 $6y^{\frac{2}{3}} + c$

9 $\dfrac{-6}{t^3} + c$

10 $6x^{\frac{5}{3}} + c$

11 $20x^{\frac{1}{4}} + c$

12 $\frac{2}{5}x^{\frac{5}{2}} + c$

13 $5x^3 + x^2 - 8x + c$

14 $\frac{4}{3}x^3 + 2x^2 + x + c$

15 $\dfrac{-3}{x} + \dfrac{1}{2}x^2 + c$

16 $\dfrac{-2}{u^2} - \dfrac{1}{u} + c$

17 $\frac{1}{6}x^6 - \frac{1}{2}x^4 + \frac{1}{3}x^3 - 2x + c$

18 $\dfrac{-3}{t} + \dfrac{1}{3}t^3 + c$

19 $2x^{0.5} + \frac{2}{3}x^{1.5} + c$

20 $\dfrac{1}{3}x^3 - \dfrac{5}{x} + c$

21 $x = 28$

22 $z = 8p^{1.5} - 54$

23 $x = 6.4$

24 $y = 5t^3 - 14t^2 + 5t + 8$

CHAPTER 5
Exercise 4

1 a) $\frac{1}{3}$ b) $\frac{8}{3}$ c) $\frac{64}{3}$ d) $\frac{125}{3}$
Area $= \frac{1}{3}b^3$

2 a) $\frac{7}{3}$ b) $\frac{26}{3}$ c) $\frac{63}{3}$ d) $\frac{124}{3}$
Area $= \frac{1}{3}b^3 - \frac{1}{3}$
$\frac{1}{3}b^3 - \frac{1}{3}$

3 a) $\frac{19}{3}$ b) $\frac{56}{3}$ c) $\frac{117}{3}$
$\frac{1}{3}b^3 - \frac{8}{3}$

4 $\frac{1}{3}b^3 - \frac{1}{3}a^3$

5 a) $\frac{1}{4}$ b) $\frac{16}{4}$ c) $\frac{81}{4}$ d) $\frac{256}{4}$
Area $= \frac{1}{4}b^4$
Area $= \frac{1}{4}b^4 - \frac{1}{4}a^4$

6 Area $= \frac{1}{5}b^5 - \frac{1}{5}a^5$

7 $\frac{1}{2}b^2 - \frac{1}{2}a^2$

CHAPTER 5
Exercise 5

1 $\frac{1}{2}$　　**2** $\frac{53}{6}$　　**3** $\frac{16}{3}$

4 $\frac{5}{6}$　　**5** $\frac{14}{3}$　　**6** $\frac{64}{5}$

7 2　　**8** 6　　**9** $\frac{5}{4}$

10 $\frac{86}{3}$　　**11** 6　　**12** 28

13 $\frac{17}{2}$　　**14** 78　　**15** $\frac{3}{4}$

16 3　　**17** $\frac{56}{5}$　　**18** 25

19 2　　**20** $p = 1.5$

CHAPTER 5
Exercise 6

1 $21\frac{1}{3}$

2 9.6

3 a) 120
　b) 30

4 16

5 51.2

6 $12\frac{2}{3}$

7 (0, 0) (8, 32)
　$42\frac{2}{3}$

8 $64\frac{4}{5}$

9 a) $y = \frac{-1}{2}x + \frac{3}{2}$　b) $Q(\frac{-3}{2}, \frac{9}{4})$　c) $\frac{125}{48}$

10 a) 60　b) $\frac{45}{4}$

11 b) 6　c) 12

CHAPTER 5
Exercise 7

1 3.13

2 8.09

3 a) π　b) 2.996

4 224 m
　2.5 m/s^2

CHAPTER 5
Revision Exercise

1 a) $6\frac{2}{3}$　b) 3

2 a) $y = \frac{1}{2}x^4 + x^3 + 2$　b) $48\frac{1}{5}$

3 $\frac{105}{2}$

4 b) (6, 18) and (−2, 4)
　c) $\frac{128}{3}$

5 b) $y = x - 1$
　d) $\frac{4}{3}$

6 a) $\frac{38}{3}$　b) $\frac{5}{4}$

7 a) $\frac{124}{5}$　b) $\frac{242}{5}$

8 21.0

9 a) 80　b) $\frac{1}{8}$

10 $\frac{1}{2}x^4 - \frac{3}{2}x^2 - \frac{6}{x} + c$

11 a) $y = 5x - x^2 - 3$　b) $(\frac{1}{2}, \frac{-3}{4})$ and (2, 3)

12 a) 4.74

CHAPTER 6
Exercise 1

9 b) $N = 20\,000 \times 1.0247^t$　c) 31 000
　d) exponential growth not possible
　indefinitely

10 b) $P = 100 \times 0.984\,84^t$　c) 47%

CHAPTER 6
Exercise 2

1 a) 3^{14}　b) p^{15}　c) 5^4
　d) 2^7　e) 7^6　f) $243x^{10}y^{20}$
　g) $\dfrac{125a^6}{8b^9}$　h) $16a^8b^{12}c^4$　i) $4x^3$
　j) $4x^2$　k) $\dfrac{1}{16x^6}$　l) $\dfrac{1}{8x^{15}}$
　m) $\dfrac{1}{81x^8}$　n) $4x^4$　o) $27x^6$
　p) $\dfrac{1}{2x^4}$　q) $\dfrac{1}{3x^4}$　r) $\dfrac{1}{25x^4}$
　s) $4x^4$　t) $\dfrac{1}{27x^6y^9}$

2 a) 3^4 b) 3^{12} c) 3^{-12}
d) 3^{-5} e) $3^{\frac{-5}{2}}$

3 a) $p = 2$ b) $p = 2$ c) $p = 5$
d) $p = -12$

CHAPTER 6
Exercise 3

1 a) 2 b) 3 c) -2
d) $\frac{1}{2}$ e) $-\frac{1}{2}$ f) 3
g) 2 h) 1 i) 0
j) -2 k) 3 l) -2
m) $\frac{1}{2}$ n) $\frac{3}{2}$ o) $-\frac{1}{2}$
p) 0 q) 2 r) $-\frac{1}{2}$
s) -2

2 a) 8 b) 32 c) $\frac{1}{5}$
d) 25 e) 3 f) $\frac{1}{4}$
g) 12 h) 16 i) $\dfrac{\sqrt{5}-1}{2}$

3 1.792

4 $y = 3^x$ can be mapped onto $y = \log_3 x$ by a reflection in the line $y = x$.

5 a) 2, 3, 5
b) $\frac{1}{2}$, 1, $\frac{3}{2}$
c) $\frac{1}{2}$, 2, $\frac{5}{2}$ $\bigg\}$ $\log_a (xy) = \log_a x + \log_a y$
d) -1, 3, 2

6 $\log_{10} x = \log_{10} 4 \times \log_4 x$

CHAPTER 6
Exercise 4

1 a) 1 b) 2 c) 2
d) 3 e) 3 f) 5
g) 1.5 h) 2 i) -1
j) 1 k) 2 l) 4.5
m) 2 n) 1

2 a) $x = 3.065$ b) $x = -0.748$
c) $x = 1.140$ d) $x = 2.161$
e) $x = 7.212$ f) $t = 1.710$

3 a) 1.404 b) 3.482
c) 3.468 d) -3.234

4 a) $x = 0$ or 2
b) $x = 0.631$ or -1
c) 1.038

5 a) $x = 1.292$ or 1.904
b) $x = 2$ or 2.807

6 a) $p + q$ b) $2p$ c) $2p + q$
d) $-2q$ e) $p - q$ f) $2p - q$
h) $2q - p$ i) $p + 3q$

7 a) $x < 7.323$ b) $n < 19.367$
c) $t > 4.187$

9 a) $n = 20.66$ b) $n = 18.14$
c) $n = 5.82$

10 $x > y$

11 a) 845.09804 ... b) 846 digits
c) 1

CHAPTER 6
Exercise 5

1 $x = \frac{1}{4}$, 2 **2** $x = 4$

3 $x = 5$ **4** $x = 5$

5 $x = 2$; $y = 8$ **6** $y = 2$; $x = 4$

CHAPTER 6
Revision Exercise

2 a) $a = 100$; $b = 25$ b) $p = 900$; $q = 675$

3 a) 3 b) -2 c) 2 d) $-\frac{1}{2}$

4 a) 2.571 b) -1.562

5 a) 4.292 b) $t \approx 1.731$
c) $y > 14.275$ d) $z < 19.367$
e) $x = 1.577$ f) $x = 1.049, -2.464$

6 a) $x = 3$ b) $x = 5$

7 a) $x = 5$ b) $x = 20$ c) $x = 8$

8 $n = 18.78$ years

9 smallest integer = 13 876

10 b) i) $N = 15\ 824$ ii) $t = 6.313$

12 a) $s = 40\ 000 \times 0.9^t$ c) $t = 31.2$
e) $t = 29.5$ f) $t = 16.8$

13 a) $x = 2$ or $-4 \pm \sqrt{19}$
c) $x = 2$ $x = -4 + \sqrt{19}$

CHAPTER 7
Exercise 1

1   15.7; 43; 79.4 $1.3n + 2.7$

2 -33; -138; -278 $-5n + 17$

3   **a)** $3n - 1$ **b)** $0.3n + 2.9$
    c) $-2n + 105$ **d)** $7n - 19$

4   **a)** $n = 1261$ **b)** $n = 40$
    c) $n = 252$

5   379 terms

6   68 terms

7 $d = 1.5$, $a - 13.5$ 71 terms

8   24, 60, 96

9 $a = 92.8$; $d = -3.2$
    29 positive terms

10   625 terms

CHAPTER 7
Exercise 2

1   **a)** 25 165.2   **b)** 10 440

2   **a)** $L = 8$ **b)** $d = 0.1$ **c)** 3.4

3   66

4   **a)** $d = 3$ **b)** 30   **c)** 630

5   **a)** 31 375   **b)** 1230   **c)** $k = 33$

6 $n = 37$

7   38 terms

8   9999 $n = 1000$ 5 503 500

9 $n = 41$

10   **a)** 500 500   **b)** 71 071
     c) 429 429   **d)** 156 361

11   **a)** 590   **b)** 58 terms needed

CHAPTER 7
Exercise 3

1   **a)** 196 608; 5.53×10^{19}; $3 \times 4^{n-1}$
    b) $\frac{1}{243}$; 1.46×10^{-14}; $27 \times \left(\frac{1}{3}\right)^{n-1}$
    c) 768; 1.29×10^{10}; $3 \times (-2)^{n-1}$

2 $n = 13$

3   **a)** $n = 10$ **b)** $n = 19$

4   8 terms

5 $a = 0.3$, $r = 2$ 1228.8

6 $r = .6389\cdots$, $a = 18.78\cdots$ 0.136

7   31 terms

8   **a)** 5   **b)** 3, 9, 15, 21, 27

CHAPTER 7
Exercise 4

1   **a)** 258 280 324   **b)** 906.64
    c) 13 655   **d)** $-27\ 305$
    e) 976 562   **f)** 12   **g)** $\frac{7}{9}$

2 $r = 2$ $a = \frac{5}{16}$ $\frac{10\ 235}{16}$

3 $r = \pm 1$, $\pm\sqrt{2}$

4   19 terms

5   **a)** 70.17 million   **b)** 37 years
    c) 600 million   **d)** $n = 25$ years

6   **a)** 100   **b)** 52 terms

7   **a)** $r = \frac{1}{2}$ or $r = \frac{1 \pm \sqrt{5}}{4}$ **b)** $r = \frac{1}{2}$ or $r = \frac{1 \pm \sqrt{5}}{4}$

8   **a)** 933.4   **b)** 80
    c) 2 097 136

9   **a)** $v_2 = 40$ $v_3 = 50$ $v_4 = 62.5$
    b) $n = 16$
    c) 8 968 182.1
    d) $N = 41$

10   **a)** £32 162.30
    b) £793 582.90
    c) during 34th year at work.

11   **a)** $\frac{37}{99}\left(1 - \left(\frac{1}{100}\right)^n\right)$ **b)** $S_n \rightarrow \frac{37}{99}$
      $5n \rightarrow \frac{37}{99}$
    c) $\frac{37}{99}$ **d)** $\frac{241}{999}$

12   **a)** 53 188.11   **b)** 28 terms

13   **a)** $200\ 000 \times 1.005^2 - 1.005P - P$
      $200\ 000 \times 1.005^3 - 1.005^2 P$
      $- 1.005P - P$
      $200\ 000 \times 1.005^4 - 1.005^3 P$
      $- 1.005^2 P - 1.005P - P$
    b) $200\ 000 \times 1.005^n - \dfrac{P(1.005^n - 1)}{0.005}$
    c) $P = \$1288.60$ \$386 580

14   As $n \rightarrow \infty$ $P_n \rightarrow \infty$ but $A_n \rightarrow \frac{9}{5}A_1$

CHAPTER 7
Revision Exercise

1   **a)** 1550   **b)** 624

2 $d = -2.4$, $a = 27.2$ 12 positive terms

3 $\frac{235}{999}$

4   6480

5   **a)** 307.2   **b)** 29 terms   **c)** 3000

6   **a)** 1.120   **b)** 99.395   **c)** 100

7   **a)** $d = 1.5$, $a = 12$
       b) 135 terms needed

9   29 terms

10 **i)** 10 700
       ii) 140 terms necessary

11 **i)** 1905
       iii) 996.6 miles

12 **a)** $v_2 = 106$ $v_3 = 112$ $v_4 = 118$
       b) $n = 152$
       c) 3140
       d) $n = 562$

CHAPTER 8
Exercise 1

1   **a)** 85.9°   **b)** 139.2°   **c)** 39.0°
       d) 4.01°   **e)** 15°   **f)** 108°

2   **a)** 0.559^c **b)** 1.855^c **c)** 3.875^c
       d) $\frac{5\pi}{12}$ **e)** $\frac{5\pi}{3}$

3   30.4 cm   57.6 cm^2

4 $\theta = 0.5^c$

5 $r = 8$ $\theta = 2$

6   16.76 m/s

CHAPTER 8
Exercise 2

1   **a)** $\theta = 1.2$ **b)** 23.29 cm
       c) 13.40 cm^2

2 $x = 4.04$ cm
       $y = 10.29$ cm
       20.04 cm^2

3 $\theta = 1.875$ radians
       57.24 cm^2

5   **a)** $\frac{\pi}{4}$, $\frac{5\pi}{4}$
       b) $\frac{7\pi}{12}$, $\frac{11\pi}{12}$, $\frac{19\pi}{12}$, $\frac{23\pi}{12}$
       c) $\frac{\pi}{6}$, $\frac{\pi}{2}$, $\frac{5\pi}{6}$, $\frac{7\pi}{6}$, $\frac{3\pi}{2}$, $\frac{11\pi}{6}$

6   **a)** $\frac{\pi}{3}$, $\frac{2\pi}{3}$, $\frac{4\pi}{3}$, $\frac{8\pi}{3}$
       b) $\frac{-11\pi}{12}$, $\frac{-7\pi}{12}$, $\frac{-\pi}{4}$, $\frac{\pi}{12}$, $\frac{5\pi}{12}$, $\frac{3\pi}{4}$
       c) $\frac{\pi}{3}$, $\frac{2\pi}{3}$, $\frac{4\pi}{3}$, $\frac{5\pi}{3}$
       d) $\frac{-2\pi}{3}$, $\frac{-\pi}{3}$, $\frac{\pi}{3}$, $\frac{2\pi}{3}$
       e) $\frac{-\pi}{2}$, $\frac{\pi}{6}$, $\frac{5\pi}{6}$

7   **a)** $\frac{-5\pi}{6}$, $\frac{-\pi}{6}$, $\frac{\pi}{2}$
       b) 0, π, 2π, $\frac{2\pi}{3}$, $\frac{4\pi}{3}$
       c) $\frac{-11\pi}{12}$, $\frac{-7\pi}{12}$, $\frac{-5\pi}{12}$, $\frac{-\pi}{12}$, $\frac{\pi}{12}$, $\frac{5\pi}{12}$, $\frac{7\pi}{12}$, $\frac{11\pi}{12}$

8   **d)** 1.30

CHAPTER 8
Revision Exercise

1   **a)** $\frac{7}{12}\pi$ **b)** $r = 5.3$ cm

2   **a)** $\frac{\pi}{3}$, $\frac{2\pi}{3}$, $\frac{4\pi}{3}$, $\frac{5\pi}{3}$
       b) $\frac{\pi}{6}$, $\frac{7\pi}{6}$

4   29.74 cm^2

5   290.3 m^2

6   **a)** $\frac{-2\pi}{3}$, $\frac{-\pi}{3}$, $\frac{\pi}{3}$, $\frac{2\pi}{3}$
       b) $\frac{-\pi}{2}$, $\frac{\pi}{2}$, $\frac{-2\pi}{3}$, $\frac{-\pi}{3}$, $\frac{\pi}{3}$, $\frac{2\pi}{3}$

7   **i)** $\theta = 0.92$ rad
       ii) 20.59 cm   16.33 cm^2

8   1.16 rad, 5.12 rad 1.91 rad, 4.37 rad.

9   **a)** $y = 2 \cos x$
       b) (0.896, 1.250)
           (2.246, −1.250)

10 a) 5π s
b) 0.72 m/s
e) $t = 4.93$ s   or   10.78 s

CHAPTER 9
Exercise 1

1   1 7 21 35 35 21 7 1
    1 8 28 56 70 56 28 8 1
    1 9 36 84 126 126 84 36 9 1

2   2, 4, 8, 16, 32, 64
    Powers of 2
    $2^{15} = 32\,768$

3   15, 21, 28
    $1 + 2 + 3 + \cdots + n = \frac{1}{2}n(n + 1)$
    Third number in each line is a triangular number.

5 $(1 + x)^2 \equiv 1 + 2x + x^2$
    $(1 + x)^3 \equiv 1 + 3x + 3x^2 + x^3$
    $(1 + x)^4 \equiv 1 + 4x + 6x^2 + 4x^3 + x^4$
    $(1 + x)^5 \equiv 1 + 5x + 10x^2 + 10x^3 + 5x^4 + x^5$
    $(1 + x)^6 \equiv 1 + 6x + 15x^2 + 20x^3 + 15x^4$
    $\qquad + 6x^5 + x^6$
    etc.

6 $(x + y)^2 \equiv x^2 + 2xy + y^2$
    $(x + y)^3 \equiv x^3 + 3x^2y + 3xy^2 + y^3$
    $(x + y)^4 \equiv x^4 + 4x^3y + 6x^2y^2 + 4xy^3 + y^4$
    etc.

CHAPTER 9
Exercise 2

1 a) $a^6 + 6a^5b + 15a^4b^2 + 20a^3b^3$
    $\qquad + 15a^2b^4 + 6ab^5 + b^6$
b) $243 + 810x + 1080x^2 + 720x^3$
    $\qquad + 240x^4 + 32x^5$
c) $125 - 300x + 240x^2 - 64x^3$

2 a) $x^3 + 12x + \dfrac{48}{x} + \dfrac{64}{x^3}$
b) $y^8 - 8y^5 + 24y^2 - \dfrac{32}{y} + \dfrac{16}{y^4}$
c) $z^6 + 6z^4 + 15z^2 + 20 + \dfrac{15}{z^2} + \dfrac{6}{z^4} + \dfrac{1}{z^6}$

3 $2187 + 5103x + 5103x^2 + 2835x^3$
    $\qquad + 945x^4 + 189x^5 + 21x^6 + x^7$
    2751.2614111

4 $32 + 152x + 252x^2 + 162x^3 + 27x^4$
    $81 - 351x + 378x^2 - 174x^3 + 37x^4 - 3x^5$

5 $3 - 25x + 90x^2 - 180x^3 + \cdots$

6 $64 - 960x + 6000x^2$
    $\alpha = 3 \qquad \beta = 4 \qquad \gamma = 14\,160$

7 $40z^8 + 1000$

9 $p = 420 \qquad 5.772$

10 a) $p^4 + 4p^3qx + 6p^2q^2x^2 + 4pq^3x^3 + q^4x^4$
b) $\alpha = 4$
    $\qquad \gamma = 5, \beta = -3$
    or $\gamma = \frac{12}{5}, \beta = \frac{89}{5}$

CHAPTER 9
Exercise 3

1 a) $1 + 16y + 120y^2 + 560y^3 + \cdots$
b) $1024 - 5120x + 11\,520x^2$
    $\qquad - 15\,360x^3 + \cdots$
c) $1 + 42x + 819x^2 + 9828x^3 + \cdots$

2 $1 + 84x + 3234x^2 + 75\,460x^3 + \cdots$
    2.238860

3 a) $144x^5$ **b)** $_{15}C_{12}(-3)^{12}x^{-6}$

4 $a = \pm\frac{1}{15} \qquad 1 - 2x + \dfrac{28}{15}x^2 - \dfrac{728}{675}x^3 + \cdots$

5   0.09602816

6 $5 - 36y + 118y^2 + \cdots$

7 a) $1024 + 1280u + 720u^2 + 240u^3$
    $\qquad + 52.5u^4 + \cdots$
b) $1024 + 1280u^2 + 720u^4 + \cdots$
c) $2048 + 1440u^4$

8 $\alpha = 3$

CHAPTER 9
Revision Exercise

1 **a)** $a^4 + 4a^3b + 6a^2b^2 + 4ab^3 + b^4$

2 **a)** $(1 + 2y)^5 = 1 + 10y + 40y^2 + 80y^3$
$+ 80y^4 + 16y^5$

b) $(1 - 2y)^5 = 1 - 10y + 40y^2 - 80y^3$
$+ 80y^4 - 16y^5$

c) $y = \pm\sqrt{7}$

3 **a)** $243 - 810x + 1080x^2 + \cdots$
b) $a = 4$

4 **a)** quotient $= 12x + 24$
b) remainder $= -59$

5 **a)** $p = 32, q = -80, r = 80$
b) -192

6 **a)** $27 + 54x^5 + 36x^{10} + 8x^{15}$
$81 + 216x^5 + 216^{10} + 96x^{15} + 16x^{20}$

7 **i)** 0.1153
ii) $1 + 36x^2 + 594x^4$
iii) 0.1132

8 50.16486 $6744\sqrt{2}$

9 $n = 7$ $p = -5$ 16 800

10 $n = 6$

REVISION 1
Trigonometry

1 **a)** $B = 14.7°$ **b)** 769.4 m

2 3.6 cm   10.8 cm^2

3 2 so

4 **a)** $x = \frac{\pi}{6}, \frac{11\pi}{6}, \frac{13\pi}{6}, \frac{23\pi}{6}$
b) $x = -75°, -15°, 105°, 165°$
c) $x = \frac{2\pi}{9}, \frac{5\pi}{9}, \frac{8\pi}{9}$

5 $x = 6$ cm $AB = 4.11$ cm

6 **a)** 4.52 radius   **b)** $r = 7.10$ cm

7 $S(-\pi, -1)$ $T(\frac{5\pi}{2}, 0)$

8 $\frac{19}{4} + 2\sqrt{3}$

9 $\theta = 9.2°, 99.2°, 189.2°, 279.2°$

10 $p = 13.3$ cm

11 $x = 367$ m

12 $(\frac{\pi}{6}, \frac{1}{2})$ $(\frac{13\pi}{6}, \frac{1}{2})$ $(\frac{5\pi}{6}, \frac{1}{2})$ $(\frac{17\pi}{6}, \frac{1}{2})$

13 **b)** $\theta = -60°, 60°,$ or $-180°, 180°$

14 $x = \sqrt{3}$ or $2\sqrt{3}$

15 $x = 23.14$ km   bearing $= 156°$

16 **a)** $\theta = 189.6, 350.4, 90°$
b) $\theta = 161.6, 341.6$

17 40.4 cm   59.5 cm^2

19 27.1 cm^2 $BC = 7.04$ cm

20 $\theta = 1.05, 5.24$ $\theta = 2.21, 4.07$

21 17 cm^2

23 $x = 0, 180, 360, 30, 150, 210, 330$

REVISION 2
Sequences

1 **i)** 64.7   **ii)** 697

2 $a = 0.5$

3 **i)** 3, 7, 3, 7, 3, 7
ii) **a)** $v_n = 10 + 3 \times (-1)^n$ **b)** 2007

4 $16x^4 + 96x^3y + 216x^2y^2 + 216xy^3 + 81xy^4$
1 609 621 621.6081

5 **a)** 200   **c)** 94 terms necessary

6 $N = 38$ $a = 17$

7 **a)** 14 950   **b)** 933.4

8 **i)** 4, 9   **ii)** $t_n + t_{n+1} = (n + 1)^2$

9 $r = 1.3$ 5428

10 **ii)** 3784   **iii)** 121

11 **i)** $a = 75$
ii) $n = 22$, 63 litres unused

12 $p = 9$ $q = 27$

13 **i)** $u_5 = (\frac{2}{3})^4 u_1 + 5[1 + (\frac{2}{3}) + (\frac{2}{3})^2 + (\frac{2}{3})^3]$
ii) $u_n = (\frac{2}{3})^{n-1} u_1 + 5[1 + (\frac{2}{3}) + (\frac{2}{3})^2 + \cdots + (\frac{2}{3})^{n-2}]$
iii) 15

14 **iii)** $p = 74$

15 $r = \frac{-2}{3}$

16 $128 + 1344x + 6048x^2 + 15\ 120x^3 + \cdots$
$128 - 1344x + 6048x^2 - 15\ 120x^3 + \cdots$
54.00192

17 i) $u_2 = \frac{3}{4}$ $u_3 = \frac{-1}{3}$ $u_4 = 4$
Sequence cycles through three values
$4, \frac{3}{4}, \frac{-1}{3}, 4, \frac{3}{4} \dots$

ii) c) -1
iii) 1752.5

18 i) $x < 0$

19 $1 - 10x + 40x^2$ $a = 2, b = 3$

20 i) $u_1 = -1$ $u_2 = -2$ $u_3 = -1$ $u_4 = -1$
oscillates between -1 and -2
ii) $u_1 = \pm \sqrt{2}$

21 i) 20 blows needed
ii) sum to infinity = 80

22 168

23 i) $u_3 = 18$ **ii)** $4n + 2$
iii) 6   4   2 004 000

24 ii) $n = 35$

REVISION 3
Algebra

1 $a = 3$

2 a) $x = 12$ **b)** $t = 4.74$

3 £3679 $n = 15$ years

4 roots are $3, \frac{2}{3}, \frac{-1}{2}$

6 -10

7 i) $0 < a < 1$ **iii)** $a = 0.9963$
iv) 186.4 days

8 $P = 1$ $Q = 3$ $R = -5$ $S = 6$

9 a) i) 3   **ii)** $\frac{-1}{2}$
b) 2   **c)** $x = 10$

10 a) $x > 6.02$ **b)** $x < 9.89$

11 a) $a = -7, b = 54$ **b)** $-3, \frac{9}{2}, 2$
c) $2x^2 + x - 17 + \dfrac{-14}{x - 4}$

12 i) $p = 25$ **ii)** $\frac{-1}{2}$ and $\frac{-3}{5}$

13 1.10×10^{6624}

14 a) 300 -50 **b)** 550
c) 250   **d)** $x = 2^{44}$

15 $x = 0.907, 1.585$

16 a) 153   **b)** $\frac{2}{3}$

17 b) $y = 32x^{-4}$

18 a) $s = 500$
b) $s = 740.1$
c) $t = 28.01$ hours

19 $x + 10 + \dfrac{22}{x - 3}$
Quotient $= x + 10$ remainder $= 22$
L_1 is $x = 3$ L_2 is $y = x + 10$

20 i) $a = 2$ **ii)** $x = 2$ or $\pm \sqrt{2}$

21 $x = 7$

22 $x = 2$ or $2 \pm \sqrt{5}$

23 a) $k = 6$
b) $3x^3 - 13x^2 - 31x + 105 = 0$

25 $x = 4$

REVISION 4
Integration

1 i) $\frac{1}{3}x^3 + \frac{1}{2}x^2 + c$ **ii)** $\dfrac{-1}{x} + c$

2 i) 14   **ii)** 19

3 a) $(-1, 6)$ and $(1.5, 3.5)$ $\frac{125}{24}$

4 $y = 2x^3 + \dfrac{4}{x^2} - 12$

5 a) 51   **b)** $a = 3$

6 144

7 6

8 -6

9 a) $P(9, 0)$

10 a) 13   **b)** $\frac{9}{4}$

12 $\frac{45}{4}$

13 a) 6   **b)** 12

14 $(0, 0)$ and $(5, 625)$ $\frac{625}{4}$

15 a) $\frac{1}{3}x^3 - \frac{4}{x} + c$ **b)** $\frac{144}{5}$

16 17.0

17 $y = 3x^2 + 4x - 2$ 11

18 $\frac{32}{3}$

19 $\frac{1}{4}x^4 - x^2 + c$ $a = 2$

20 4.38   under-estimate   17.5

21 $y = \frac{-1}{2}x + \frac{3}{2}$ $\frac{125}{48}$

22 **i)** $\frac{5}{4}$ m   **ii)** 17.6 m^3

Sample exam paper

1 **i)** $16 + 32x + 24x^2 + 8x^3 + x^4$

2 $6\frac{2}{3}$

3 **ii)** 75 750

4 **i)** 1.97 rad   **ii)** 18.9 cm^2

5 **b)** 6   **ii)** $a = 4$

6 **i)** $k = 16$ **ii)** $x = 3$, $x = \frac{1}{2}$, $x = \frac{1}{2}$

7 **a)** $5\sqrt{3}$ **b)** $\tan \theta$
    c) 0, 48.2, 180

8 **i)** $n = 17$ **ii)** $p = 2$, $q = 100$

9 **a)** $x = \frac{1}{k} \log_2 \left(1 - \frac{y}{a}\right)$ **b)** 5, –20   **c)** 1

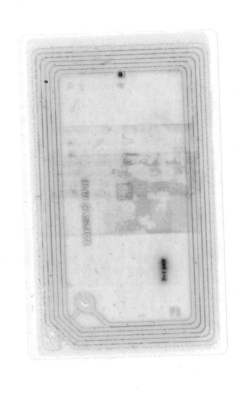